# Student Lab Manual

## for

# Argument-Driven Inquiry

## in

# PHYSICAL SCIENCE

## LAB INVESTIGATIONS
## for GRADES 6–8

# Student Lab Manual

## for

# Argument-Driven Inquiry

## in

# PHYSICAL SCIENCE

## LAB INVESTIGATIONS
## for GRADES 6–8

Jonathon Grooms, Patrick J. Enderle, Todd Hutner,
Ashley Murphy, and Victor Sampson

NSTApress
National Science Teachers Association
Arlington, Virginia

National Science Teachers Association

Claire Reinburg, Director
Wendy Rubin, Managing Editor
Rachel Ledbetter, Associate Editor
Amanda Van Beuren, Associate Editor
Donna Yudkin, Book Acquisitions Coordinator

**ART AND DESIGN**
Will Thomas Jr., Director

**PRINTING AND PRODUCTION**
Catherine Lorrain, Director

**NATIONAL SCIENCE TEACHERS ASSOCIATION**
David L. Evans, Executive Director
David Beacom, Publisher

1840 Wilson Blvd., Arlington, VA 22201
*www.nsta.org/store*
For customer service inquiries, please call 800-277-5300.

*NSTA is committed to publishing material that promotes the best in inquiry-based science education. However, conditions of actual use may vary, and the safety procedures and practices described in this book are intended to serve only as a guide. Additional precautionary measures may be required. NSTA and the authors do not warrant or represent that the procedures and practices in this book meet any safety code or standard of federal, state, or local regulations. NSTA and the authors disclaim any liability for personal injury or damage to property arising out of or relating to the use of this book, including any of the recommendations, instructions, or materials contained therein.*

**Cataloging-in-Publication Data are available from the Library of Congress.**
LCCN: 2016027981
ISBN: 978-1-68140-526-1

# CONTENTS

# SECTION 3—Physical Science Core Idea 2
## Motion and Stability: Forces and Interactions

## INTRODUCTION LABS

## APPLICATION LABS

# SECTION 4—Physical Science Core Idea 3
## Energy

## INTRODUCTION LABS

## APPLICATION LABS

## SECTION 5—Physical Science Core Idea 4
## Waves and Their Applications in Technologies for Information Transfer

## INTRODUCTION LABS

## APPLICATION LABS

# ACKNOWLEDGMENTS

The development of this book was supported by the Institute of Education Sciences, U.S. Department of Education, through grant R305A100909 to Florida State University. The opinions expressed are those of the authors and do not represent the views of the institute or the U.S. Department of Education.

X

# ABOUT THE AUTHORS

**Jonathon Grooms** is an assistant professor of curriculum and pedagogy in the Graduate School of Education and Human Development at The George Washington University. He received a BS in secondary science and mathematics teaching with a focus in chemistry and physics from Florida State University (FSU). Upon graduation, Jonathon joined FSU's Office of Science Teaching, where he directed the physical science outreach program "Science on the Move." He also earned a PhD in science education from FSU. His research interests include student engagement in scientific argumentation and students' application of argumentation strategies in socioscientific contexts. To learn more about his work in science education, go to *www.jgrooms.com*.

**Patrick J. Enderle** is an assistant professor of science education in the Department of Middle and Secondary Education at Georgia State University. He received his BS and MS in molecular biology from East Carolina University. Patrick spent some time as a high school biology teacher and several years as a visiting professor in the Department of Biology at East Carolina University. He then attended FSU, from which he graduated with a PhD in science education. His research interests include argumentation in the science classroom, science teacher professional development, and enhancing undergraduate science education. To learn more about his work in science education, go to *http://patrickenderle.weebly.com*.

**Todd Hutner** is a research associate in the Center for STEM Education (see *http://stemcenter.utexas.edu*) at The University of Texas at Austin (UT-Austin). He received a BS and an MS in science education from FSU and a PhD in curriculum and instruction from UT-Austin. Todd also taught high school physics and chemistry for four years. He specializes in teacher learning, teacher practice, and educational policy in science education.

**Ashley Murphy** attended FSU and earned a BS with dual majors in biology and secondary science education. Ashley spent some time as a middle school biology and science teacher before entering graduate school at UT-Austin, where she is currently working toward a PhD in STEM (science, technology, engineering, and mathematics) education. Her research interests include argumentation in middle and elementary classrooms. As an educator, she frequently employed argumentation as a means to enhance student understanding of concepts and science literacy.

**Victor Sampson** is an associate professor of STEM education and the director of the Center for STEM Education at UT-Austin. He received a BA in zoology from the University of Washington, an MIT from Seattle University, and a PhD in curriculum and instruction with a specialization in science education from Arizona State University. Victor also taught high school biology and chemistry for nine years. He specializes in argumentation in science education, teacher learning, and assessment. To learn more about his work in science education, go to *www.vicsampson.com*.

# SECTION 1
# Introduction and Lab Safety

# INTRODUCTION

## By Patrick J. Enderle and Victor Sampson

Science is more than a collection of facts or ideas that describe what we know about how the world works and why it works that way. Science is also a set of crosscutting concepts and practices that people can use to develop and refine new explanations for, or descriptions of, the natural world. When you understand the core ideas, crosscutting concepts, and practices of science, it is easier to appreciate the beauty and wonder of science, to engage in public discussions about science, and to evaluate the strengths of scientific information presented through popular media. You will also have the knowledge and skills needed to continue learning science outside school or to enter a career in science, engineering, or technology.

The core ideas of science include the theories, laws, and models that scientists use to explain natural events or bodies of data and to predict the results of new investigations. The crosscutting concepts are themes that have value in all areas of science and are used to help us understand many different natural phenomena. They can be used to connect knowledge from the various fields of science into a coherent and scientifically based view of the world and to help us think about what might be important to consider or look for during an investigation. Finally, the practices of science are used to develop and refine new ideas about the world. Although some practices are specific to certain fields of science, all fields share a set of common practices. The practices include such things as asking and answering questions; planning and carrying out investigations; analyzing and interpreting data; and obtaining, evaluating, and communicating information. One of the most important practices of science is arguing from evidence. Arguing from evidence involves proposing, supporting, challenging, and refining claims based on evidence. Arguing is important because scientists need to be able to examine, review, and evaluate their own ideas and to critique those of others. Scientists also argue from evidence when they need to judge the quality of data, produce and improve models, develop new questions from those models that can be investigated, and suggest ways to refine or modify existing theories, laws, and models.

Always remember that science is a social activity, not an individual one. Science is social because many different scientists contribute to the development of new scientific knowledge. As scientists carry out their research, they frequently talk with their colleagues, both formally, like at a meeting, and informally, like in a hallway. They exchange emails, engage in discussions at conferences, share research techniques and analytical procedures, and present new ideas by writing articles in journals or chapters in books. They also critique the ideas and methods used by other scientists through a formal peer review process before those ideas can be published in journals or books. In short, scientists are members of a community, the members of which work together to build, develop, test, critique, and refine ideas. The ways scientists talk, write, think, and interact with each other reflect common ideas about what counts as quality and shared standards for how new ideas should be developed, shared, evaluated, and refined. These ways of interacting make science different from other ways of knowing. The core ideas, crosscutting concepts, and practices of science are important among scientists because most, if not all, scientists find them to be useful in developing and refining new explanations for, or descriptions of, the natural world.

The laboratory investigations included in this book are designed to help you learn the core ideas, crosscutting concepts, and practices of science. During each investigation, you will have an opportunity to use a core idea, several crosscutting concepts, and the practices of science to understand a natural phenomenon or solve a problem. Your teacher will introduce each investigation by giving you a task to accomplish and a guiding question to answer. You will then work as part of a team to plan and carry out an investigation to collect the data you need to answer that question. From there, your team will develop an initial argument that includes a claim, evidence in support of your claim, and a justification of your evidence. The claim will be your answer to the guiding question, the evidence will include your analysis of the data you collected and an interpretation of that analysis, and the justification will explain why your evidence is important in terms of key science concepts. Next, you will have an opportunity to share your argument with your classmates and to critique their arguments, much like professional scientists do. You will then revise your initial argument based on your colleagues' feedback. Finally, you will be asked to write an investigation report on your own to share what you learned. The report will go through a double-blind peer review so you can improve it before you submit it to you teacher for a grade. As you complete more and more investigations in this lab manual, you will not only learn the core ideas associated with each investigation but also get better at using the crosscutting concepts and practices of science to understand the natural world.

# SAFETY IN THE SCIENCE CLASSROOM, LABORATORY, OR FIELD SITES

*Note to science teachers and supervisors/administrators: The following safety acknowledgment form is for your use in the classroom and should be given to students at the beginning of the school year to help them understand their role in ensuring a safer and productive science experience.*

Science is a process of discovering and exploring the natural world. Exploration occurs in the classroom/laboratory or in the field. As part of your science class, you will be doing many activities and investigations that will involve the use of various materials, equipment, and chemicals. Safety in the science classroom, laboratory, or field sites is the FIRST PRIORITY for students, instructors, and parents. To ensure safer classroom/laboratory/field experiences, the following **Science Rules and Regulations** have been developed for the protection and safety of all. Your instructor will provide additional rules for specific situations or settings. The rules and regulations must be followed at all times. After you have reviewed them with your instructor, read and review the rules and regulations with your parent/guardian. Their signature and your signature on the safety acknowledgment form are required before you will be permitted to participate in any activities or investigations. Your signature indicates that you have read these rules and regulations, understand them, and agree to follow them at all times while working in the classroom/laboratory or in the field.

## Safety Standards of Student Conduct in the Classroom, Laboratory, and in the Field

1. Conduct yourself in a responsible manner at all times. Frivolous activities, mischievous behavior, throwing items, and conducting pranks are prohibited.

2. Lab and safety information and procedures must be read ahead of time. All verbal and written instructions shall be followed in carrying out the activity or investigation.

3. Eating, drinking, gum chewing, applying cosmetics, manipulating contact lenses, and other unsafe activities are not permitted in the laboratory.

4. Working in the laboratory without the instructor present is prohibited.

5. Unauthorized activities or investigations are prohibited. Unsupervised work is not permitted.

6. Entering preparation or chemical storage areas is prohibited at all times.

7. Removing chemicals or equipment from the classroom or laboratory is prohibited unless authorized by the instructor.

## Personal Safety

8. Sanitized indirectly vented chemical splash goggles or safety glasses as appropriate (meeting the ANSI Z87.1 standard) shall be worn during activities or demonstrations in the classroom, laboratory, or field, including pre-laboratory work and clean-up, unless the instructor specifically states that the activity or demonstration does not require the use of eye protection.

9. When an activity requires the use of laboratory aprons, the apron shall be appropriate to the size of the student and the hazard associated with the activity or investigation. The apron shall remain tied throughout the activity or investigation.

10. All accidents, chemical spills, and injuries must be reported immediately to the instructor, no matter how trivial they may seem at the time. Follow your instructor's directions for immediate treatment.

11. Dress appropriately for laboratory work by protecting your body with clothing and shoes. This means that you should use hair ties to tie back long hair and tuck into the collar. Do not wear loose or baggy clothing or dangling jewelry on laboratory days. Acrylic nails are also a safety hazard near heat sources and should not be used. Sandals or open-toed shoes are not to be worn during any lab activities. Refer to pre-lab instructions. If in doubt, ask!

12. Know the location of all safety equipment in the room. This includes eye wash stations, the deluge shower, fire extinguishers, the fume hood, and the safety blanket. Know the location of emergency master electric and gas shut offs and exits.

13. Certain classrooms may have living organisms including plants in aquaria or other containers. Students must not handle organisms without specific instructor authorization. Wash your hands with soap and water after handling organisms and plants.

14. When an activity or investigation requires the use of laboratory gloves for hand protection, the gloves shall be appropriate for the hazard and worn throughout the activity.

## Specific Safety Precautions Involving Chemicals and Lab Equipment

15. Avoid inhaling in fumes that may be generated during an activity or investigation.

16. Never fill pipettes by mouth suction. Always use the suction bulbs or pumps.

17. Do not force glass tubing into rubber stoppers. Use glycerin as a lubricant and hold the tubing with a towel as you ease the glass into the stopper.

18. Proper procedures shall be followed when using any heating or flame producing device especially gas burners. Never leave a flame unattended.

19. Remember that hot glass looks the same as cold glass. After heating, glass remains hot for a very long time. Determine if an object is hot by placing your hand close to the object but do not touch it.

20. Should a fire drill, lockdown, or other emergency occur during an investigation or activity, make sure you turn off all gas burners and electrical equipment. During an evacuation emergency, exit the room as directed. During a lockdown, move out of the line of sight from doors and windows if possible or as directed.

21. Always read the reagent bottle labels twice before you use the reagent. Be certain the chemical you use is the correct one.

22. Replace the top on any reagent bottle as soon as you have finished using it and return the reagent to the designated location.

23. Do not return unused chemicals to the reagent container. Follow the instructor's directions for the storage or disposal of these materials.

## Standards For Maintaining a Safer Laboratory Environment

24. Backpacks and books are to remain in an area designated by the instructor and shall not be brought into the laboratory area.

25. Never sit on laboratory tables.

26. Work areas should be kept clean and neat at all times. Work surfaces are to be cleaned at the end of each laboratory or activity.

27. Solid chemicals, metals, matches, filter papers, broken glass, and other materials designated by the instructor are to be deposited in the proper waste containers, not in the sink. Follow your instructor's directions for disposal of waste.

28. Sinks are to be used for the disposal of water and those solutions designated by the instructor. Other solutions must be placed in the designated waste disposal containers.

29. Glassware is to be washed with hot, soapy water and scrubbed with the appropriate type and sized brush, rinsed, dried, and returned to its original location.

30. Goggles are to be worn during the activity or investigation, clean up, and through hand washing.

31. Safety Data Sheets (SDSs) contain critical information about hazardous chemicals of which students need to be aware. Your instructor will review the salient points on the SDSs for the hazardous chemicals students will be working with and also post the SDSs in the lab for future reference.

## Safety Acknowledgment Form: Science Rules and Regulations

I have read the science rules and regulations in the *Student Lab Manual for Argument-Driven Inquiry in Life Science,* and I agree to follow them during any science course, investigation, or activity. By signing this form, I acknowledge that the science classroom, laboratory, or field sites can be an unsafe place to work and learn. The safety rules and regulations are developed to help prevent accidents and to ensure my own safety and the safety of my fellow students. I will follow any additional instructions given by my instructor. I understand that I may ask my instructor at any time about the rules and regulations if they are not clear to me. My failure to follow these science laboratory rules and regulations may result in disciplinary action.

_____          _____
Student Signature                                      Date

_____          _____
Parent/Guardian Signature                         Date

# SECTION 2
# Physical Sciences Core Idea 1

## Matter and Its Interactions

# Introduction Labs

# LAB 1

## Lab 1. Thermal Energy and Matter
### What Happens at the Molecular Level When Thermal Energy Is Added to a Substance?

### Introduction

Every substance in the universe is made up of matter. A substance can exist in three different states: solid, liquid, or gas. A substance such as water can easily transition from one state of matter to the other. For example, water transitions from a solid state to a liquid state when an ice cube melts (Figure L1.1). The ice cube is able to melt and transition from a solid to a liquid because it absorbs thermal energy. Thermal energy is a type of energy that is transferred between two objects because they have different temperatures. In the example of an ice cube melting, thermal energy is transferred to the ice cube from the warm air surrounding it. Thermal energy always moves from the warmer object to the colder object. Think about another example, such as a cold can of soda in your hand. In that case, thermal energy is transferring from your hand to the soda; eventually the cold soda will gain enough thermal energy that it becomes the same temperature as its surroundings.

### FIGURE L1.1

**Water undergoes a phase change from solid to liquid when ice melts.**

All substances, regardless of whether they are a solid, a liquid, or a gas, are made up of atoms and molecules. Atoms and molecules are submicroscopic, meaning they are too small to be seen with our eyes and even too small to be seen with most microscopes. The atoms or molecules that make up a substance are constantly in motion. The composition of a substance will always be the same even though the substance can transition from one state to a different state. Water, for example, is always made of $H_2O$ molecules even when it is in a solid (ice), liquid, or gaseous (steam) state.

The difference between the solid, liquid, and gas states of a substance is due to the amount of kinetic energy the atoms or molecules have and how these particles are moving relative to each other. Kinetic energy is the energy of motion. Atoms or molecules that are moving quickly have more kinetic energy than atoms or molecules that move more slowly. For example, the molecules found within a sample of gaseous water move around quickly. These molecules therefore have a lot of kinetic energy. The molecules that are found in a sample of solid water, in contrast, move around slower and have less kinetic energy than the molecules in a sample of gaseous water. You can therefore measure the temperature of a substance to learn about the average kinetic energy of the molecules within that substance.

At this point, we have established several key ideas about the nature of matter. For example, we know that all matter can exist in three different states and all matter is composed of atoms or molecules that are really small. We also know that a substance has the same composition regardless of its state and that the atoms or molecule of a substance will have different amounts of kinetic energy at different temperatures. These ideas, when taken together, can serve as a foundation for the development of an explanatory model that can be used to illustrate what happens at the molecular level when thermal energy is added to a substance. This type of model is important to develop because explanatory models can help us predict the behavior of matter under different conditions. For example, we could use an explanatory model to help us predict how long it will take a substance to boil when it is heated on a hot plate or a stove. Your goal for this investigation will be to collect data about the behavior of a substance when thermal energy is added to it and then use what you learn to develop an explanatory model that describes what happens to the molecules that make up a substance when they are exposed to thermal energy.

## Your Task

Develop a model that helps you explain what happens at the molecular level as thermal energy is added to a substance. The substance you will work with during this investigation is water. Your model should account for the mass of the substance, its temperature, and the amount of time that thermal energy is being added to the substance so that you can explain the relationship between the amount of water in a sample, the temperature at which the sample boils, and how long it takes to reach the boiling temperature. Your model, once fully developed, should enable you to make accurate predictions about the amount of time it will take for a particular sample of water to boil. Once you have developed your model, you will need to test it to determine if it leads to accurate predictions or not.

The guiding question of this investigation is, **What happens at the molecular level when thermal energy is added to a substance?**

## Materials

You may use any of the following materials during your investigation:

| Consumable | Equipment |
|---|---|
| • Water | • Beakers (various sizes) |
| | • Graduated cylinders (various sizes) |
| | • Electronic or triple beam balance |
| | • Hot plate |
| | • Thermometer or temperature probe |
| | • Stopwatch |
| | • Safety glasses or goggles |
| | • Chemical-resistant apron |
| | • Nonlatex gloves |

## Safety Precautions

Follow all normal lab safety rules. In addition, take the following safety precautions:

1. Wear sanitized indirectly vented chemical-splash goggles and chemical-resistant nonlatex gloves and aprons during lab setup, hands-on activity, and takedown.

2. Never put consumables in your mouth.

3. Use caution when working with hot plates, because they can burn skin and cause fires.

4. Hot plates also need to be kept away from water and other liquids.

5. Use only GFCI-protected electrical receptacles for hot plates.

6. Clean up any spilled water immediately to avoid a slip or fall hazard.

7. Be careful when working with hot water, because it can burn skin.

8. Handle all glassware with care.

9. Handle glass thermometers with care. They are fragile and can break, causing a sharp hazard that can cut or puncture skin.

10. Never return the consumables to stock bottles.

11. Wash hands with soap and water after completing the lab activity.

**Investigation Proposal Required?**   ☐ Yes      ☐ No

## FIGURE L1.2

**A sample of water can be heated on a hot plate.**

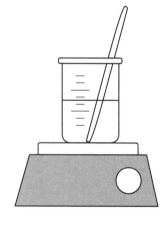

## Getting Started

The first step in developing your model is to design and carry out an investigation to determine how long it takes for different samples of water to boil and the temperature at which each sample boils. To accomplish this task, you can heat a sample of water in a beaker on a hot plate (see Figure L1.2). Before you begin to heat different samples of water, you must determine what type of data you need to collect, how you will collect it, and how you will analyze it.

To determine *what type of data you need to collect*, think about the following questions:

• What information do you need to make your model?

• What measurements will you take during your investigation?

• How will you know how much thermal energy has been transferred to your samples of water?

To determine *how you will collect the data*, think about the following questions:

• What equipment will you use to collect the data you need?

• How will you make sure that your data are of high quality (i.e., how will you reduce error)?

• How will you keep track of the data you collect?

- How will you organize your data?

To determine *how you will analyze the data*, think about the following questions:

- What type of calculations will you need to make?
- What type of table or graph could you create to help make sense of your data?

Once you have carried out your investigations, your group will need to develop a model that can be used to help explain what is happening at the molecular level when thermal energy is added to water. Your model must include the relationship between the amount of water being heated, the temperature at which that sample of water boils, and how long it takes the sample to reach the boiling temperature. Your model should also be able to account for any differences in the mass of water, differences in initial temperature between samples, and the amount of time that thermal energy is added to the substance.

The last step in this investigation is to test your model. To accomplish this goal, you can heat different amounts of water that you did not investigate to determine if your model leads to accurate predictions about the time it takes for each particular sample of water to boil. If you are able to use your model to make accurate predictions about the time it takes for different amounts of water to boil, then you will be able to generate the evidence you need to convince others that the model you developed is valid.

## Connections to Crosscutting Concepts, the Nature of Science, and the Nature of Scientific Inquiry

As you work through your investigation, be sure to think about

- how scientists need to be able to recognize what is relevant at different scales;
- how scientists often need to track how energy moves into, out of, and within a system;
- the difference between laws and theories in science; and
- how scientists must use imagination and creativity when developing models and explanations.

## Initial Argument

Once your group has finished collecting and analyzing your data, your group will need to develop an initial argument. Your argument needs to include a *claim*, *evidence* to support your claim, and a *justification* of the evidence. The claim is your group's answer to the guiding question. The evidence is an analysis and interpretation of your data. Finally, the justification of the evidence is why your group thinks the evidence matters. The justification of the evidence is important because scientists can use different kinds of evidence to support their claims. Your group will create your initial argument on a whiteboard.

# LAB 1

**Argument presentation on a whiteboard**

| The Guiding Question: | |
|---|---|
| Our Claim: | |
| Our Evidence: | Our Justification of the Evidence: |

Your whiteboard should include all the information shown in Figure L1.3.

## Argumentation Session

The argumentation session allows all of the groups to share their arguments. One member of each group will stay at the lab station to share that group's argument, while the other members of the group go to the other lab stations to listen to and critique the arguments developed by their classmates. This is similar to how scientists present their arguments to other scientists at conferences. If you are responsible for critiquing your classmates' arguments, your goal is to look for mistakes so these mistakes can be fixed and they can make their argument better. The argumentation session is also a good time to think about ways you can make your initial argument better. Scientists must share and critique arguments like this to develop new ideas.

To critique an argument, you might need more information than what is included on the whiteboard. You will therefore need to ask the presenter lots of questions. Here are some good questions to ask:

- How did you collect your data? Why did you use that method? Why did you collect those data?

- What did you do to make sure the data you collected are reliable? What did you do to decrease measurement error?

- How did your group analyze the data? Why did you decide to do it that way? Did you check your calculations?

- Is that the only way to interpret the results of your analysis? How do you know that your interpretation of your analysis is appropriate?

- Why did your group decide to present your evidence in that way?

- What other claims did your group discuss before you decided on that one? Why did your group abandon those alternative ideas?

- How confident are you that your claim is valid? What could you do to increase your confidence?

Once the argumentation session is complete, you will have a chance to meet with your group and revise your initial argument. Your group might need to gather more data or design a way to test one or more alternative claims as part of this process. Remember, your goal at this stage of the investigation is to develop the most acceptable and valid answer to the research question!

## Report

Once you have completed your research, you will need to prepare an *investigation report* that consists of three sections. Each section should provide an answer to the following questions:

1. What question were you trying to answer and why?

2. What did you do to answer your question and why?

3. What is your argument?

Your report should answer these questions in two pages or less. The report must be typed, and any diagrams, figures, or tables should be embedded into the document. Be sure to write in a persuasive style; you are trying to convince others that your claim is acceptable or valid!

# LAB 1

# Lab 1. Thermal Energy and Matter

## What Happens at the Molecular Level When Thermal Energy Is Added to a Substance?

The image below shows three beakers filled with water. Each beaker is sitting on a different hot plate, and each hot plate is set to a specific temperature. Draw a model inside each circle that explains the behavior of the water molecules in each beaker.

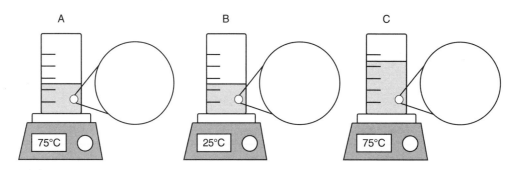

1. Explain your model. Why did you draw it that way?

2. A scientist has two beakers of water. As shown in the figure at right, one beaker has 50 ml of water at 75°C and the other beaker has 75 ml of water at 25°C. She then mixes the water together in a third beaker.

   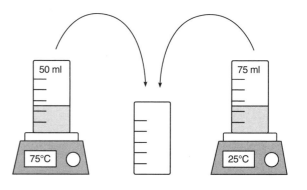

   What will be the temperature of the water after they are mixed?

   a. > 45°C

   b. 45°C

   c. < 45°C

   d. Unsure

Explain your answer. What rule did you use to make your decision?

3. A scientific law describes the behavior of a natural phenomenon or a generalized relationship under certain conditions; a scientific theory is a well-substantiated explanation.

   a. I agree with this statement.
   b. I disagree with this statement.

   Explain your answer, using an example from your investigation about thermal energy.

4. Scientists need to be creative and have a good imagination.

   a. I agree with this statement.
   b. I disagree with this statement.

   Explain your answer, using an example from your investigation about thermal energy.

5. Scientists often need to be able to recognize proportional relationships between groups or quantities. Explain why recognizing a proportional relationship is important, using an example from your investigation about thermal energy.

6. It is often important to track how matter flows into, out of, and within a system during an investigation. Explain why it is important to keep track of matter when studying a system, using an example from your investigation about thermal energy.

*Lab Handout*

# Lab 2. Chemical and Physical Changes
## What Set of Rules Should We Use to Distinguish Between Chemical and Physical Changes in Matter?

### Introduction

Matter has mass and occupies space. Although these two basic characteristics of matter are shared among all the different kinds of matter in the universe, most kinds have even more characteristics that are special to only certain groups or types. Scientists work to understand the special characteristics of the many kinds of matter so that they can also understand what happens when different kinds of matter interact with each other.

Every kind of matter, such as wood, steel, or water, has a unique set of physical and chemical properties. Physical properties of matter include qualities or attributes such as density, odor, color, melting point, boiling point, and magnetism. These properties are often useful for identifying different types of substances. Chemical properties of matter, in contrast, describe how matter interacts with other types of matter. For example, when a metal is added to an acid, it reacts with the acid to form a gas. A chemical property of metal, therefore, is reactivity with acids. All substances have a specific set of physical and chemical properties. Scientists can therefore use physical and chemical properties to identify an unknown substance.

The matter that is around us changes all of the time. Natural events, for example, can change matter in a variety of ways, such as when a forest fire turns wood into ash or when wind breaks down a rock to produce small particles of dust. Another example of a change in matter is when a solid turns into a liquid, which is what happens when a crayon melts (see Figure L2.1). Scientists classify changes in matter as either a chemical change or a physical change. A chemical change is defined as a change in the composition and properties of a substance. Chemical changes involve the rearrangement of molecules or atoms and result in the production of one or more new substances. A physical change is defined as a change in the state or energy of matter. A physical change does not result in the production of a new substance, although the starting and ending materials may look very different from each other.

Matter will look very different after it goes through a chemical change because a chemical change transforms a substance into one or more new substances. The appearance of matter, however, can also change when it goes through a physical change. In addition, we cannot see what happens at the level of atoms as matter goes through a change. It is

### FIGURE L2.1

**A crayon melting**

therefore often difficult to tell the difference between a chemical change and a physical change by just observing the appearance of a substance. Scientists, as a result, have developed several rules to help them classify changes in matter. In this investigation, you will have an opportunity to develop a set of rules that you can use to determine if a change in matter should be classified as a chemical change or a physical change.

## Your Task

Use what you know about physical and chemical properties of matter, stability and change, and how to design and carry out an investigation to develop a set of rules you can use to distinguish between a chemical change and a physical change in matter.

The guiding question of this investigation is, **What set of rules should we use to distinguish between chemical and physical changes in matter?**

## Materials

You may use any of the following materials during your investigation:

| Consumables | Equipment |
|---|---|
| • Water | • Well plate |
| • Baking soda | • Beakers (various sizes) |
| • Calcium chloride | • Electronic or triple beam balance |
| • Sugar | • Graduated cylinders (various sizes) |
| • Vinegar | |
| • Antacid tablets | • Pipettes |
| • Steel wool | • Spatulas |
| • Paper | • Matches |
| • Candle | • Mortar and pestle |
| | • pH paper |
| | • Conductivity probe |
| | • Safety glasses or goggles |
| | • Chemical-resistant apron |
| | • Nonlatex gloves |

## Safety Precautions

Follow all normal lab safety rules. Your teacher will provide important information about working with the chemicals associated with this investigation. In addition, take the following safety precautions:

1. Follow safety precautions noted on safety data sheets for hazardous chemicals.

2. Wear sanitized indirectly vented chemical-splash goggles and chemical-resistant nonlatex gloves and aprons during lab setup, hands-on activity, and takedown.

3. Never put consumables in your mouth.

4. Use caution when working with hot plates or candles. They can burn skin and cause fires.

5. Hot plates also need to be kept away from water and other liquids.

6. Use only GFCI-protected electrical receptacles for hot plates.

7. Clean up any spilled liquid immediately to avoid a slip or fall hazard.

8. Use caution when working with hazardous chemicals in this lab that are corrosive and/or toxic.

9. Handle all glassware with care.

10. Use caution in handling steel wool, because it can cause metal slivers.

11. Never return the consumables to stock bottles.

12. Wash hands with soap and water after completing the lab activity.

## Investigation Proposal Required?     ☐ Yes          ☐ No

## Getting Started

The first step in this investigation is to identify all the various physical properties of matter that are possible to observe or measure using the available materials. Once you know what physical properties you can observe or measure during this investigation, you can then start collecting data about what happens to the physical properties of matter when it goes through a physical or a chemical change. Be sure to observe or measure several different physical properties of the matter you are using before and after the change takes place. Listed below are some examples of actions to cause physical and chemical changes that you can use to document the physical properties of the matter before and after it goes through a change.

*Actions to cause a physical change*

- Grind an antacid tablet.
- Cut a piece of steel wool to its smallest size.
- Put 1 g of candle wax in a test tube and heat the test tube in a hot-water bath.
- Add 1 g of sugar to 5 ml water.

*Actions to cause a chemical change*

- Add 1 g of baking soda to 5 ml of vinegar.
- Add 1 g of steel wool to 5 ml of vinegar.
- Add 1 g of antacid tablet to 5 ml water.
- Add 1 g of sugar in a test tube and then heat it using a candle.

# LAB 2

The second step in this investigation is to develop a set of rules that you can use to determine if a change in matter is a physical one or a chemical one. Once you have a set of rules, you will need to test them to determine if they allow you to accurately identify a physical or chemical change in matter. It is important for you to test your rules because the results of your test will allow you to demonstrate that your rules are not only valid but also a useful way to identify a change in matter. Be sure to modify your rules as needed if they do not allow you to accurately classify a change in matter. To accomplish this final step of your investigation, you can test your rules using the following examples of actions to cause physical and chemical changes:

*Actions to cause a physical change*

- Add 1 ml of water to a balloon and tie the balloon so the water inside the balloon cannot escape. Then place the balloon in a microwave and heat the water (in the balloon) for 30 seconds.
- Mix 1 g of calcium chloride and 1 g of baking soda.
- Mix 1 g of crushed antacid and 1 g of sugar.

*Actions to cause a chemical change*

- Add 1 g of calcium chloride to 5 ml of water.
- Add 1 g of baking soda to 5 ml of water.
- Add 1 g of calcium chloride and 1 g of baking soda to 5 ml of water.

Remember, if you can use your rules to classify these changes in matter correctly, then you will be able to generate the evidence you need to convince others that the rules that you developed are valid and useful.

## Connections to Crosscutting Concepts, the Nature of Science, and the Nature of Scientific Inquiry

As you work through your investigation, be sure to think about

- the importance of identifying patterns,
- stability and change in nature,
- the difference between observations and inferences, and
- the role of imagination and creativity in science.

## Initial Argument

Once your group has finished collecting and analyzing your data, your group will need to develop an initial argument. Your argument needs to include a *claim*, *evidence* to support your claim, and a *justification* of the evidence. The claim is your group's answer to the guiding question. The evidence is an analysis and interpretation of your data. Finally, the justification of the evidence is why your group thinks the evidence matters. The justification of the evidence is important because scientists can use different kinds of evidence to support their claims. Your group will create your initial argument on a whiteboard. Your whiteboard should include all the information shown in Figure L2.2.

## FIGURE L2.2

**Argument presentation on a whiteboard**

| The Guiding Question: | |
|---|---|
| Our Claim: | |
| Our Evidence: | Our Justification of the Evidence: |

## Argumentation Session

The argumentation session allows all of the groups to share their arguments. One member of each group will stay at the lab station to share that group's argument, while the other members of the group go to the other lab stations to listen to and critique the arguments developed by their classmates. This is similar to how scientists present their arguments to other scientists at conferences. If you are responsible for critiquing your classmates' arguments, your goal is to look for mistakes so these mistakes can be fixed and they can make their argument better. The argumentation session is also a good time to think about ways you can make your initial argument better. Scientists must share and critique arguments like this to develop new ideas.

To critique an argument, you might need more information than what is included on the whiteboard. You will therefore need to ask the presenter lots of questions. Here are some good questions to ask:

- How did you collect your data? Why did you use that method? Why did you collect those data?

- What did you do to make sure the data you collected are reliable? What did you do to decrease measurement error?

- How did your group analyze the data? Why did you decide to do it that way? Did you check your calculations?

- Is that the only way to interpret the results of your analysis? How do you know that your interpretation of your analysis is appropriate?

- Why did your group decide to present your evidence in that way?

- What other claims did your group discuss before you decided on that one? Why did your group abandon those alternative ideas?

- How confident are you that your claim is valid? What could you do to increase your confidence?

Once the argumentation session is complete, you will have a chance to meet with your group and revise your initial argument. Your group might need to gather more data or design a way to test one or more alternative claims as part of this process. Remember, your goal at this stage of the investigation is to develop the most acceptable and valid answer to the research question!

## Report

Once you have completed your research, you will need to prepare an *investigation report* that consists of three sections. Each section should provide an answer to the following questions:

1. What question were you trying to answer and why?

2. What did you do to answer your question and why?

3. What is your argument?

Your report should answer these questions in two pages or less. This report must be typed, and any diagrams, figures, or tables should be embedded into the document. Be sure to write in a persuasive style; you are trying to convince others that your claim is acceptable and valid!

*Checkout Questions*

# Lab 2. Chemical and Physical Changes
## What Set of Rules Should We Use to Distinguish Between Chemical and Physical Changes in Matter?

1. What is a physical change in matter?

2. What is a chemical change in matter?

3. A scientist has a collection of substances, both solids and liquids, that she mixes in different combinations. For each mixture, she puts a smaller amount of solid into a larger amount of liquid. She is trying to determine if mixing these substances produces any physical or chemical changes.

| Mixture | Solid | Liquid | Observation after mixing |
|---|---|---|---|
| A | Sugar—white crystals | Water—clear | Clear solution |
| B | Sugar—white crystals | Vinegar—clear | Clear solution |
| C | Baking soda—white powder | Water—clear | Clear solution |
| D | Baking soda—white powder | Vinegar—clear | Clear solution with bubbles |

a. Which mixture(s) involve only a physical change?

b. How do you know?

# LAB 2

c.  Which mixture(s) involve only a chemical change?

d.  How do you know?

4.  Scientists do not use creativity or imagination when they are investigating the physical world.

   a.  I agree with this statement.
   b.  I disagree with this statement.

   Explain your answer, using an example from your investigation about physical and chemical changes.

5.  The result of mixing vinegar and baking soda is a chemical change.

   a.  I agree with this statement.
   b.  I disagree with this statement.

   Explain your answer, using an example from your investigation about physical and chemical changes.

6. Scientists often need to look for patterns that occur in the data they collect and analyze. Explain why identifying patterns is important, using an example from your investigation about physical and chemical changes.

7. Physical systems will become stable over time after experiencing a period of change. Explain why it is important to understand how a system stabilizes after experiencing change, using an example from your investigation about physical and chemical changes.

# Application Labs

## Lab Handout

# Lab 3. Physical Properties of Matter
## What Are the Identities of the Unknown Substances?

### Introduction

Matter, the "stuff" of which the universe is composed, is all around us. Anything that we can touch, feel, or see is an example of matter. Matter can be defined as something that has mass and takes up space. All matter is composed of submicroscopic particles called atoms. A substance is a sample of matter that has a constant composition. Examples of substances include water, iron, plastic, and glass. On Earth, substances are found in one of three different states (i.e., solid, liquid, and gas), and it is common to see a substance change from one state to another. The types of atoms, the interactions that occur between atoms, and how the atoms are moving within a substance determine its state and its behavior under different conditions.

Scientists use atomic composition and specific chemical or physical properties to distinguish between different substances (see Figure L3.1). The atomic composition of a substance refers to the different types of atoms found in it and the relative proportion of each type of atom. Water, for example, is composed of hydrogen atoms and oxygen atoms in a ratio of two hydrogen atoms for every one oxygen atom. The chemical and physical properties of a substance refer to measurable or observable qualities or attributes that are used to distinguish between different substances. Chemical properties describe how a substance interacts with other matter. Sodium and potassium, for example, react with water, but aluminum and gold do not. Physical properties are descriptive characteristics of matter. Examples of physical properties include color, density, conductivity, and malleability. Every substance will have a unique set of chemical and physical properties that can be used to identify it, because every type of substance has a unique atomic composition.

It is often challenging to determine the identity of an unknown substance based on its chemical and physical properties. A scientist, for example, may only have a small amount of a substance. As a result, the scientist may not be able to conduct all the different types of tests that he or she wants to because some tests may change the characteristics of the sample during the process (such as when a metal is mixed with an acid). It is also difficult to determine many of the physical properties of the sample, such as its density or its malleability, when there is only a small amount of the substance, because taking

## FIGURE L3.1

**How scientists distinguish between different substances**

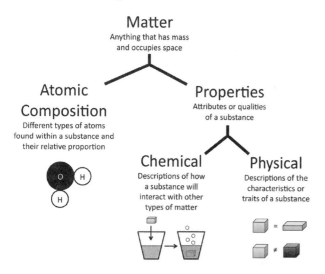

measurements is harder. To complicate matters further, an unknown substance may have an irregular shape, which can make it difficult to accurately measure its volume. Without knowing the mass and the volume of a substance, it is impossible to calculate its density.

In this investigation, you will have an opportunity to learn about some of the challenges scientists face when they need to identify an unknown substance based on its physical properties and why it is important to make accurate measurements inside the laboratory.

## Your Task

You will be given a set of known substances. You will then document, measure, or calculate at least three different physical properties for each substance. From there, you will return the known substances to your teacher, who will then give you a set of unknown substances. The unknown substances will consist of one or more of the known substances. Your goal is to use what you know about the physical properties of matter, proportional relationships, and patterns to design and carry out an investigation that will enable you to collect the data you need to determine the identity of the unknown substances.

The guiding question of this investigation is, **What are the identities of the unknown substances?**

## Materials

You may use any of the following materials during your investigation:

**Consumables**
- Water (in squirt bottles)
- Set of known substances
- Set of unknown substances

**Equipment**
- Electronic or triple beam balance
- Beakers (various sizes)
- Graduated cylinders (various sizes)
- Pipettes
- Metric ruler
- Wire
- Size D battery
- Mini lightbulb
- Mini lightbulb holder
- Safety glasses or goggles
- Chemical-resistant apron
- Nonlatex gloves

## Safety Precautions

Follow all normal lab safety rules. In addition, take the following safety precautions:

1. Wear sanitized indirectly vented chemical-splash goggles and chemical-resistant nonlatex gloves and aprons during lab setup, hands-on activity, and takedown.

2. Clean up any spilled liquid immediately to avoid a slip or fall hazard.

3. Handle all glassware with care.

4. Lightbulbs are made of glass. Be careful handling them. If they break, clean them up immediately and place in a broken glass box.

5. Handle electrical wires with caution. They have sharp ends, which can cut or puncture skin.

6. Wash hands with soap and water after completing the lab activity.

## Investigation Proposal Required?     ☐ Yes     ☐ No

## Getting Started

To answer the guiding question, you will need to make several systematic observations of the known and unknown substances. To accomplish this task, you must determine what type of data you need to collect, how you will collect it, and how you will analyze it.

To determine *what type of data you need to collect,* think about the following questions:

- Which three physical properties (e.g., color, density, conductivity, malleability, luster) will you focus on as you make your systematic observations?
- What information do you need to determine or calculate each of the physical properties?

To determine *how you will collect the data,* think about the following questions:

- What equipment will you need to collect the data you need?
- How will you make sure that your data are of high quality (i.e., how will you reduce error)?
- How will you keep track of the data you collect?
- How will you organize your data?

To determine *how you will analyze the data,* think about the following questions:

- What type of calculations will you need to make?
- What patterns do you need to look for in your data?
- What type of table or graph could you create to help make sense of your data?
- How will you determine if the physical properties of the various objects are the same or different?

### Connections to Crosscutting Concepts, the Nature of Science, and the Nature of Scientific Inquiry

As you work through your investigation, be sure to think about

- how scientists use patterns as a basis for classification systems;
- how scientists need to be able to recognize proportional relationships between categories, groups, or quantities;
- the difference between data and evidence in science; and
- how scientists use different types of methods to answer different types of questions.

### Initial Argument

Once your group has finished collecting and analyzing your data, your group will need to develop an initial argument. Your initial argument needs to include a *claim*, *evidence* to support your claim, and a *justification* of the evidence. The claim is your group's answer to the guiding question. The evidence is an analysis and interpretation of your data. Finally, the justification of the evidence is why your group thinks the evidence matters. The justification of the evidence is important because scientists can use different kinds of evidence to support their claims. Your group will create your initial argument on a whiteboard. Your whiteboard should include all the information shown in Figure L3.2.

## FIGURE L3.2

**Argument presentation on a whiteboard**

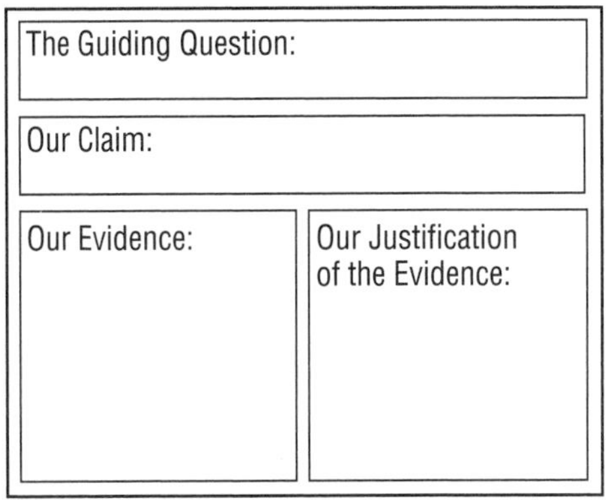

| The Guiding Question: | |
|---|---|
| Our Claim: | |
| Our Evidence: | Our Justification of the Evidence: |

### Argumentation Session

The argumentation session allows all of the groups to share their arguments. One member of each group will stay at the lab station to share that group's argument, while the other members of the group go to the other lab stations to listen to and critique the arguments developed by their classmates. This is similar to how scientists present their arguments to other scientists at conferences. If you are responsible for critiquing your classmates' arguments, your goal is to look for mistakes so these mistakes can be fixed and they can make their argument better. The argumentation session is also a good time to think about ways you can make your initial argument better. Scientists must share and critique arguments like this to develop new ideas.

To critique an argument, you might need more information than what is included on the whiteboard. You will therefore need to ask the presenter lots of questions. Here are some good questions to ask:

- How did you collect your data? Why did you use that method? Why did you collect those data?

- What did you do to make sure the data you collected are reliable? What did you do to decrease measurement error?

- How did your group analyze the data? Why did you decide to do it that way? Did you check your calculations?

- Is that the only way to interpret the results of your analysis? How do you know that your interpretation of your analysis is appropriate?

- Why did your group decide to present your evidence in that way?

- What other claims did your group discuss before you decided on that one? Why did your group abandon those alternative ideas?

- How confident are you that your claim is valid? What could you do to increase your confidence?

Once the argumentation session is complete, you will have a chance to meet with your group and revise your initial argument. Your group might need to gather more data or design a way to test one or more alternative claims as part of this process. Remember, your goal at this stage of the investigation is to develop the most acceptable and valid answer to the research question!

## Report

Once you have completed your research, you will need to prepare an *investigation report* that consists of three sections. Each section should provide an answer to the following questions:

1. What question were you trying to answer and why?

2. What did you do to answer your question and why?

3. What is your argument?

Your report should answer these questions in two pages or less. This report must be typed, and any diagrams, figures, or tables should be embedded into the document. Be sure to write in a persuasive style; you are trying to convince others that your claim is acceptable and valid!

*Checkout Questions*

# Lab 3. Physical Properties of Matter
## What Are the Identities of the Unknown Substances?

1. What is the difference between a physical and a chemical property of matter?

2. Why do substances have unique physical and chemical properties?

3. A scientist is given four different objects. She collects the following information about them.

| Object | Mass | Volume | Conducts electricity? | Malleable? |
|--------|--------|----------|-----------------------|------------|
| A | 24.5 g | 2.5 cm³ | Yes | Yes |
| B | 10.1 g | 1.5 cm³ | Yes | No |
| C | 4.6 g | 2.2 cm³ | No | No |
| D | 27.7 g | 3.8 cm³ | Yes | Yes |

a. Are any of the objects the same substance?

b. How do you know?

4. Scientists can use an experiment to answer any type of research question.

    a. I agree with this statement.

    b. I disagree with this statement.

Explain your answer, using an example from your investigation about identification of an unknown based on physical properties.

5. "The freezing points of solutions A and B are both –1°C" is an example of evidence.

    a. I agree with this statement.

    b. I disagree with this statement.

Explain your answer, using an example from your investigation about identification of an unknown based on physical properties.

6. Scientists often need to look for proportional relationships between different quantities during an investigation. Explain what a proportional relationship is and why these relationships are important, using an example from your investigation about identification of an unknown based on physical properties.

7. It is often important for scientists to identify patterns during an investigation. Explain why it is important to identify patterns when studying a system, using an example from your investigation about identification of an unknown based on physical properties.

# Lab 4. Conservation of Mass

## How Does the Total Mass of the Substances Formed as a Result of a Chemical Change Compare With the Total Mass of the Original Substances?

### Introduction

Matter is defined as anything that has mass and takes up space. Matter is composed of submicroscopic particles called atoms. To date, we know of 118 different types of atoms. All atoms share the same basic structure. At the center of an atom is a nucleus, which is composed of even smaller particles called protons and neutrons. Atoms are also composed of a third type of particle called electrons, which are found in specific regions around the nucleus. These regions are called orbitals. Scientists use the number of protons found in the nucleus of an atom to distinguish between the 118 different types of atoms. For example, there is 1 proton in the nucleus of a hydrogen atom and 30 protons in the nucleus of a zinc atom. Each type of atom also has a specific mass that reflects the composition of its nucleus.

Atoms can be bonded together in different combinations to create different types of molecules. Atoms or molecules can be combined to create different types of substances. A substance is a sample of matter that has a constant composition. Substances that consist of a single type of atom, such as gold or tin, are called elements. Substances that consist of a single type of molecule, such as water or sugar, are called compounds. A substance has qualities or attributes that distinguish it from other substances. These qualities or attributes are called physical and chemical properties. Physical properties are observable or measurable characteristics of a substance. Examples of physical properties include such things as density, melting point, and boiling point. Chemical properties, in contrast, describe how a substance interacts with other substances. For example, zinc reacts with hydrochloric acid but not with water. Scientists can identify a substance by examining its physical and chemical properties because every type of substance has a unique set of physical and chemical properties that reflect its unique atomic or molecular composition.

A substance can go through both chemical and physical changes. Any change in a substance that involves a rearrangement of how the atoms within that substance are bonded together is called a chemical change. A chemical change causes one or more substances to be transformed into one or more different substances. This process is often described as a chemical reaction. The original substance or substances involved in the chemical reaction are called reactants and the new substance or substances are called products. A physical change in matter, in contrast, does not involve a rearrangement of how the atoms within that substance are bonded together. A physical change is simply a change in the appearance of a substance. Examples of a physical change include a liquid turning into a solid or a solid turning into a liquid and a substance being broken or cut into smaller pieces. Figure L4.1 illustrates what happens at the submicroscopic level when a

# Conservation of Mass

*How Does the Total Mass of the Substances Formed as a Result of a Chemical Change Compare With the Total Mass of the Original Substances?*

## FIGURE L4.1

**Difference between a chemical change and physical change at the submicroscopic level**

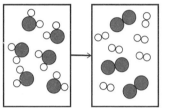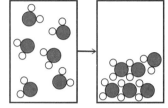

substance, such as water, goes through a chemical or physical change.

Many substances will react with other substances in predictable ways. Take the reaction of zinc and hydrochloric acid as an example. When zinc and hydrochloric acid are mixed together, the resulting products of the reaction will be hydrogen gas and a solution of zinc chloride. Another example is the reaction that takes place between a solution of silver nitrate and a solution of sodium chloride. When these two clear solutions are mixed together, the atoms in each substance interact and then rearrange to produce a different solution containing sodium nitrate and a solid substance called silver chloride. The properties of the products that are formed as a result of a chemical reaction are different than the properties of the reactants because the atoms in the original substances were broken apart and then rearranged and combined in a new way. The new configuration of atoms results in substances that have a different atomic or molecular composition. The unique atomic or molecular composition of a substance, as noted earlier, gives a substance its unique chemical and physical properties.

The chemical and physical properties of the reactants and the products of a chemical reaction are often very different even when the reactants and the products are composed of the same types of atoms. To illustrate, consider what happens when zinc (a metal) and hydrochloric acid (HCl molecules dissolved in water) are mixed. Zinc and hydrochloric acid, as noted earlier, react to produce hydrogen (a gas) and a solution of zinc chloride ($ZnCl_2$ molecules dissolved in water). Table L4.1 shows the composition of these four substances and some of the physical properties of each one. As can be seen in the table, zinc and hydrochloric acid (the reactants) have very different physical properties than hydrogen and zinc chloride (the products), even though the reactants and the products of this reaction are composed of the same three types of atoms.

## TABLE L4.1

**Formulas and some physical properties of zinc, hydrochloric acid, hydrogen, and zinc chloride**

| Substance | Formula | Physical properties | | | |
|-----------|---------|---------------------|---|---|---|
| | | Density (g/cm³) | Phase (at 23°C) | Melting point (°C) | Boiling point (°C) |
| **Zinc** | Zn | 7.14 | Solid | 419 | 907 |
| **Hydrochloric acid** | HCl | 1.2 | Liquid | −26 | 48 |
| **Hydrogen** | $H_2$ | 0.00009 | Gas | −259 | −253 |
| **Zinc chloride** | $ZnCl_2$ | 2.9 | Solid | 290 | 732 |

# LAB 4

At this point, we have established several fundamental ideas about the nature of matter. We know that all matter has mass, that matter is composed of atoms, and that each type of atom has a specific mass. We also know that the reactants and the products of a reaction contain the same types of atoms, because a chemical change is just a rearrangement of atoms. These fundamental ideas, when taken together, suggest that the total mass of the reactants should be the same as the total mass of the products left at the end of a chemical reaction. This claim, however, seems highly unlikely, because the substances that are left at the end of a reaction often have very different physical properties than the substances at the start of the reaction. Your goal for this investigation will be to test the validity or the acceptability of this hypothesis.

## Your Task

Use what you to know about atoms, chemical reactions, systems, and how to track the movement of matter to design and carry out an investigation to determine if the total mass of the reactants is same as the total mass of the products left at the end of a chemical reaction.

The guiding question of this investigation is, **How does the total mass of the substances formed as a result of a chemical change compare with the total mass of the original substances?**

## Materials

You may use any of the following materials during your investigation:

**Consumables**
- Sodium bicarbonate, $NaHCO_3$
- Magnesium (Mg) metal ribbon
- 1 M acetic acid, $C_2H_4O_2$
- 1 M hydrochloric acid, HCl
- 0.1 M aluminum nitrate, $Al(NO_3)_3$
- 0.1 M sodium hydroxide, NaOH
- 0.1 M copper(II) nitrate, $Cu(NO_3)_2$

**Equipment**
- 4 Beakers (various sizes)
- 4 Erlenmeyer flasks (various sizes)
- 2 5.0 ml Test tubes
- 4 Rubber stoppers
- 4 Balloons
- Weighing dishes or paper
- Electronic or triple beam balance
- Safety glasses or goggles
- Chemical-resistant apron
- Nonlatex gloves

## Safety Precautions

Follow all normal lab safety rules. Acetic acid, hydrochloric acid, and sodium hydroxide are corrosive to eyes, skin, and other body tissues. Aluminum nitrate, copper(II) nitrate, and sodium hydroxide are toxic by ingestion. Your teacher will explain relevant and important information about working with the chemicals associated with this investigation. In addition, take the following safety precautions:

1. Follow safety precautions noted on safety data sheets for hazardous chemicals.

# Conservation of Mass

*How Does the Total Mass of the Substances Formed as a Result of a Chemical Change Compare With the Total Mass of the Original Substances?*

2. Wear sanitized indirectly vented chemical-splash goggles and chemical-resistant nonlatex gloves and aprons during lab setup, hands-on activity, and takedown.

3. Never put consumables in your mouth.

4. Clean up any spilled liquid immediately to avoid a slip or fall hazard.

5. Use caution when working with hazardous chemicals in this lab that are corrosive and/or toxic.

6. Handle all glassware with care.

7. Never return the consumables to stock bottles.

8. Follow proper procedure for disposal of chemicals and solutions.

9. Wash hands with soap and water after completing the lab activity.

## Investigation Proposal Required?      ☐ Yes          ☐ No

## Getting Started

To answer the guiding question, you will investigate four different chemical reactions. The reactants and products for each chemical reaction are provided in Table L4.2. Your goal is to determine if the total mass of the reactants that you use in each reaction is the same or different than the total mass of the products.

## TABLE L4.2

**Reactants and products of the four chemical reactions**

| Reaction | Reactants | Products |
|---|---|---|
| 1 | Sodium bicarbonate (s) and acetic acid (aq) | Carbon dioxide (g), sodium acetate (aq), and water (l) |
| 2 | Magnesium (s) and hydrochloric acid (aq) | Magnesium chloride (aq) and hydrogen (g) |
| 3 | Aluminum nitrate (aq) and sodium hydroxide (aq) | Aluminum hydroxide (s) and sodium nitrate (aq) |
| 4 | Copper(II) nitrate (aq) and sodium hydroxide (aq) | Copper hydroxide (s) and sodium nitrate (aq) |

*Note:* aq = aqueous solution (solid dissolved in water); g = gas ; l = liquid; s = solid.

Some of products that you will produce during your investigation will be solids, some will be liquids, and some will be gases. Your challenge will be to find a way to ensure that none of the substances that you create when you mix the reactants together escape from the container you are using to hold them during the reaction or once the reaction is complete. You will only be given a limited amount of each reactant, so it is important to find a way to create a closed system before you mix any of the reactants together. You will also need to determine what type of data you need to collect, how you will collect it, and how you will analyze it before you begin your investigation.

To determine *what type of data you need to collect,* think about the following questions:

- What observations (color change, production of gas, etc.) will you need to make during your investigation?
- What measurements (mass of the reactants, mass of the containers, etc.) will you need to make during your investigation?

To determine *how you will collect the data,* think about the following questions:

- How will you ensure that none of the substances that you create when you mix the reactants together escape during the reaction or once the reaction is complete?
- How will you take into account the mass of the containers?
- When will you need to make your observations or measurements?
- What equipment will you need to collect the data?
- How will you make sure that your data are of high quality (i.e., how will you reduce error)?
- How will you keep track of the data you collect?
- How will you organize your data?

To determine *how you will analyze the data,* think about the following questions:

- What type of calculations will you need to make?
- What type of table or graph could you create to help make sense of your data?
- How will you determine if the total mass of the reactants and the products is the same or different?

## Connections to Crosscutting Concepts, the Nature of Science, and the Nature of Scientific Inquiry

As you work through your investigation, be sure to think about

- the importance of defining a system under study;
- how scientists often need track how matter moves into, out of, and within a system;
- the difference between data and evidence in science; and
- how testing explanations requires imagination and creativity.

# Conservation of Mass

*How Does the Total Mass of the Substances Formed as a Result of a Chemical Change Compare With the Total Mass of the Original Substances?*

## Initial Argument

Once your group has finished collecting and analyzing your data, your group will need to develop an initial argument. Your initial argument needs to include a *claim, evidence* to support your claim, and a *justification* of the evidence. The claim is your group's answer to the guiding question. The evidence is an analysis and interpretation of your data. Finally, the justification of the evidence is why your group thinks the evidence matters. The justification of the evidence is important because scientists can use different kinds of evidence to support their claims. Your group will create your initial argument on a whiteboard. Your whiteboard should include all the information shown in Figure L4.2.

## FIGURE L4.2

**Argument presentation on a whiteboard**

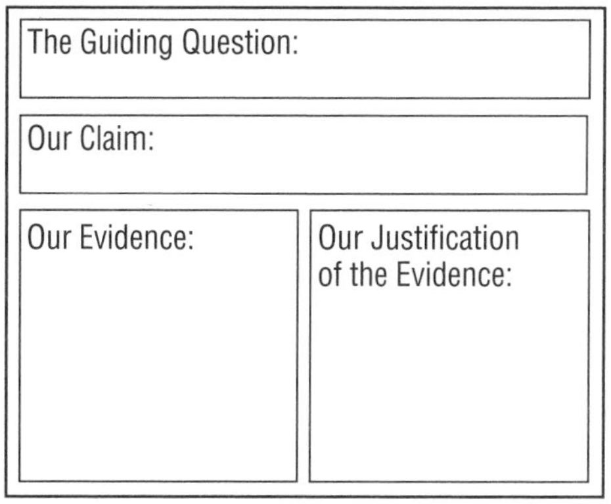

| The Guiding Question: | |
|---|---|
| Our Claim: | |
| Our Evidence: | Our Justification of the Evidence: |

## Argumentation Session

The argumentation session allows all of the groups to share their arguments. One member of each group will stay at the lab station to share that group's argument, while the other members of the group go to the other lab stations to listen to and critique the arguments developed by their classmates. This is similar to how scientists present their arguments to other scientists at conferences. If you are responsible for critiquing your classmates' arguments, your goal is to look for mistakes so these mistakes can be fixed and they can make their argument better. The argumentation session is also a good time to think about ways you can make your initial argument better. Scientists must share and critique arguments like this to develop new ideas.

To critique an argument, you might need more information than what is included on the whiteboard. You will therefore need to ask the presenter lots of questions. Here are some good questions to ask:

- How did you collect your data? Why did you use that method? Why did you collect those data?

- What did you do to make sure the data you collected are reliable? What did you do to decrease measurement error?

- How did your group analyze the data? Why did you decide to do it that way? Did you check your calculations?

- Is that the only way to interpret the results of your analysis? How do you know that your interpretation of your analysis is appropriate?

- Why did your group decide to present your evidence in that way?

- What other claims did your group discuss before you decided on that one? Why did your group abandon those alternative ideas?

• How confident are you that your claim is valid? What could you do to increase your confidence?

Once the argumentation session is complete, you will have a chance to meet with your group and revise your initial argument. Your group might need to gather more data or design a way to test one or more alternative claims as part of this process. Remember, your goal at this stage of the investigation is to develop the most acceptable and valid answer to the research question!

## Report

Once you have completed your research, you will need to prepare an *investigation report* that consists of three sections. Each section should provide an answer to the following questions:

1. What question were you trying to answer and why?

2. What did you do to answer your question and why?

3. What is your argument?

Your report should answer these questions in two pages or less. This report must be typed, and any diagrams, figures, or tables should be embedded into the document. Be sure to write in a persuasive style; you are trying to convince others that your claim is acceptable and valid!

*Checkout Questions*

# Lab 4. Conservation of Mass

## How Does the Total Mass of the Substances Formed as a Result of a Chemical Change Compare With the Total Mass of the Original Substances?

1. The figure below shows a submicroscopic view of matter going through either a physical or chemical change.

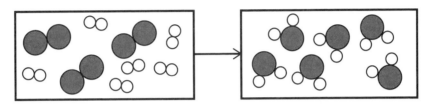

   What type of change is illustrated in this figure?

   a. A physical change

   b. A chemical change

   c. Unsure

   Explain your answer. What rule did you use to make your decision?

2. A scientist takes the mass of two liquids. As shown in the figure below, the total mass of the two liquids and the containers holding them is 245 g. She then mixes the two liquids and a chemical reaction take place. A precipitate is produced as a result of the reaction.

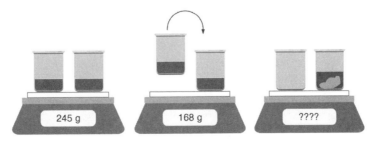

What will be the total mass of the products and the containers holding them?

a.  > 245 g
b.  245 g
c.  < 245 g
d.  Unsure

Explain your answer. What rule did you use to make your decision?

3.  Scientists do not need to have a good imagination or to be creative when testing ideas.

a.  I agree with this statement.
b.  I disagree with this statement.

Explain your answer, using an example from your investigation about the conservation of mass.

# Conservation of Mass

*How Does the Total Mass of the Substances Formed as a Result of a Chemical Change Compare With the Total Mass of the Original Substances?*

4. "The mass of the reactants is 245 grams" is an example of evidence.

    a. I agree with this statement.

    b. I disagree with this statement.

    Explain your answer, using an example from your investigation about the conservation of mass.

5. Scientists often need to define the system under study during an investigation. Explain why defining the system under study is important, using an example from your investigation about the conservation of mass.

6. It is often important to track how matter flows into, out of, and within a system during an investigation. Explain why it is important to keep track of matter when studying a system, using an example from your investigation about the conservation of mass.

## Lab Handout

# Lab 5. Design Challenge
## Which Design Will Cool a Soda the Best?

### Introduction

Many chemical processes are accompanied by a change in temperature of the substances involved. These temperature changes occur because energy is either released or absorbed during the process. When energy is released during a chemical process, the process is exothermic. Think about a hand warmer pack that you might place in a glove on a cold day. A chemical reaction is taking place inside the pack, which releases energy that is absorbed by your hand and warms them up. Other chemical processes need to absorb energy to occur; these processes are endothermic. Cold packs inside a first aid kit are a good example of an endothermic process. The pack is normally at room temperature, but shaking it up activates it, and the chemicals inside are allowed to mix together and they get very cold. The cold pack feels cold because it must absorb energy from its surroundings to help the chemicals dissolve together. The energy the cold pack absorbs is coming from your hand and the air, and because energy is leaving your hand, your hand gets cold.

Endothermic and exothermic processes are common, but sometimes it is necessary to control the flow of energy during these processes. Substances called insulators slow down the transfer of heat energy. Insulation in the walls and ceiling of homes is used to keep a home warm for longer on a cold winter day. The same insulation also helps keep the inside of the home cool on a hot summer day. Some materials work very well as insulators, but others do not. Styrofoam is a good insulator and is often used to keep hot drinks warm, such as coffee (see Figure L5.1), but it is also used to keep other drinks cold, such as soda. Engineers often have to investigate how much insulation to use for different products so that they can regulate the amount of thermal energy that transfers into or out of a system, so they can design a quality product.

## FIGURE L5.1

**Cardboard sleeves on coffee cups act as insulators to slow the transfer of energy from the hot beverage to our hand.**

### Your Task

Use what you know about chemical reactions, thermal energy, and how energy flows into and out of a system to design a product that will help to cool a can of soda (or other beverage) by using a chemical reaction. The structure and function of your design should be related in a way that causes the drink to cool down but also ensures that your hand does not get cold if you are holding the drink.

The guiding question of this investigation is, **Which design will cool a soda the best?**

## Materials

You may use any of the following materials during your investigation:

**Consumables**
- Ammonium chloride, $NH_4Cl$
- Ammonium nitrate, $NH_4NO_3$
- Calcium chloride, $CaCl_2$
- Magnesium sulfate, $MgSO_4$
- Sodium bicarbonate, $NaHCO_3$
- Sodium chloride, $NaCl$
- Sodium thiosulfate, $Na_2S_2O_3$
- Paper cups
- Plastic baggies
- Water or soda

**Equipment**
- Polystyrene cups
- Beaker (50 ml)
- Graduated cylinder (25 ml)
- Insulators
- Electronic or triple beam balance
- Thermometer or temperature probe
- Chemical spatula
- Scissors
- Tape
- Rubber bands
- Safety glasses or goggles
- Chemical-resistant apron
- Nonlatex gloves

## Safety Precautions

Follow all normal lab safety rules. Ammonium chloride, ammonium nitrate, sodium thiosulfate, and magnesium sulfate are all moderately toxic by ingestion and tissue irritants. Your teacher will explain relevant and important information about working with the chemicals associated with this investigation.

1. Follow safety precautions noted on safety data sheets for hazardous chemicals.

2. Wear sanitized indirectly vented chemical-splash goggles and chemical-resistant nonlatex gloves and aprons during lab setup, hands-on activity, and takedown.

3. Never put consumables in your mouth.

4. Clean up any spilled liquid immediately to avoid a slip or fall hazard.

5. Use caution when working with hazardous chemicals in this lab that are corrosive, toxic, and/or irritant.

6. Handle all glassware with care.

7. Handle glass thermometers with care. They are fragile and can break, causing a sharp hazard that can cut or puncture skin.

8. Never return the consumables to stock bottles.

9. Follow proper procedure for disposal of chemicals and solutions.

10. Wash hands with soap and water after completing the lab activity.

## Investigation Proposal Required?    ☐ Yes        ☐ No

## Getting Started

To answer the guiding question, you will need to make systematic observations of the various salts. To accomplish this task, you must determine what type of data you need to collect, how you will collect it, and how you will analyze it.

To determine *what type of data you need to collect*, think about the following questions:

- How will you know if a chemical will help make the drink sample cold?
- What information do you need to determine if your design is successful?

To determine *how you will collect your data*, think about the following questions:

- What equipment will you use to collect the data you need?
- How will you make sure that your data are of high quality (i.e., how will you reduce error)?
- How will you keep track of the data you collect?
- How will you organize your data?

To determine *how you will analyze your data*, think about the following questions:

- What type of calculations will you need to make?
- What type of table or graph could you create to help make sense of your data?
- How will you determine if your design is successful or not?

## Connections to Crosscutting Concepts, the Nature of Science, and the Nature of Scientific Inquiry

As you work through your investigation, be sure to think about

- how scientists often need to track how energy moves into, out of, and within a system;
- the relationship between the structure and function of an object;
- how scientists and engineers use a variety of methods to answer questions and solve problems; and
- the role of imagination and creativity when scientists and engineers attempt to solve problems.

# LAB 5

## Initial Argument

Once your group has finished collecting and analyzing your data, your group will need to develop an initial argument. Your initial argument needs to include a *claim*, *evidence* to support your claim, and a *justification* of the evidence. The claim is your group's answer to the guiding question. The evidence is an analysis and interpretation of your data. Finally, the justification of the evidence is why your group thinks the evidence matters. The justification of the evidence is important because scientists can use different kinds of evidence to support their claims. Your group will create your initial argument on a whiteboard. Your whiteboard should include all the information shown in Figure L5.2.

## FIGURE L5.2

**Argument presentation on a whiteboard**

| The Guiding Question: | |
| --- | --- |
| Our Claim: | |
| Our Evidence: | Our Justification of the Evidence: |

## Argumentation Session

The argumentation session allows all of the groups to share their arguments. One member of each group will stay at the lab station to share that group's argument, while the other members of the group go to the other lab stations to listen to and critique the arguments developed by their classmates. This is similar to how scientists present their arguments to other scientists at conferences. If you are responsible for critiquing your classmates' arguments, your goal is to look for mistakes so these mistakes can be fixed and they can make their argument better. The argumentation session is also a good time to think about ways you can make your initial argument better. Scientists must share and critique arguments like this to develop new ideas.

To critique an argument, you might need more information than what is included on the whiteboard. You will therefore need to ask the presenter lots of questions. Here are some good questions to ask:

- How did you collect your data? Why did you use that method? Why did you collect those data?
- What did you do to make sure the data you collected are reliable? What did you do to decrease measurement error?
- How did your group analyze the data? Why did you decide to do it that way? Did you check your calculations?
- Is that the only way to interpret the results of your analysis? How do you know that your interpretation of your analysis is appropriate?
- Why did your group decide to present your evidence in that way?
- What other claims did your group discuss before you decided on that one? Why did your group abandon those alternative ideas?

- How confident are you that your claim is valid? What could you do to increase your confidence?

Once the argumentation session is complete, you will have a chance to meet with your group and revise your initial argument. Your group might need to gather more data or design a way to test one or more alternative claims as part of this process. Remember, your goal at this stage of the investigation is to develop the most acceptable and valid answer to the research question!

## Report

Once you have completed your research, you will need to prepare an *investigation report* that consists of three sections. Each section should provide an answer to the following questions:

1. What question were you trying to answer and why?

2. What did you do to answer your question and why?

3. What is your argument?

Your report should answer these questions in two pages or less. This report must be typed, and any diagrams, figures, or tables should be embedded into the document. Be sure to write in a persuasive style; you are trying to convince others that your claim is acceptable and valid!

# LAB 5

## Lab 5. Design Challenge
### Which Design Will Cool a Soda the Best?

1. Describe what you know about materials that act as insulators and conductors when it comes to heat energy.

2. Insulated coffee mugs are popular because they keep coffee warm for much longer than a simple paper cup. Vacuum-style containers, such as the one shown below, are made with empty space between the walls.

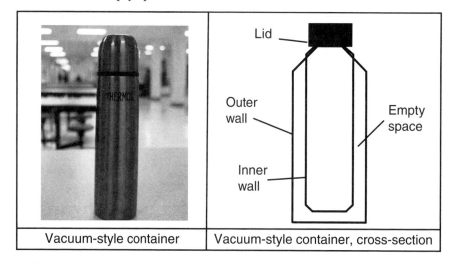

| Vacuum-style container | Vacuum-style container, cross-section |

Use what you know about insulators and conductors to explain why this style of container is good for keeping liquids hot or cold.

3. Scientists use the same procedures to answer all the questions they investigate in their work.

   a.  I agree with this statement.
   b.  I disagree with this statement.

   Explain your answer, using an example from your investigation about thermal energy.

4. Scientists do not need to be creative and have a good imagination.

   a.  I agree with this statement.
   b.  I disagree with this statement.

   Explain your answer, using an example from your investigation about thermal energy.

5. Scientists often study the various relationships between the structure of an object and its functions. Explain why recognizing these relationships is important, using an example from your investigation about thermal energy.

6. It is often important to track how matter flows into, out of, and within system during an investigation. Explain why it is important to keep track of matter when studying a system, using an example from your investigation about thermal energy.

# SECTION 3
# Physical Science
# Core Idea 2

## Motion and Stability:
## Forces and Interactions

# Introduction Labs

## Lab Handout

# Lab 6. Strength of Gravitational Force
## How Does the Gravitational Force That Exists Between Two Objects Relate to Their Masses and the Distance Between Them?

### Introduction

The motion of an object is the result of all the different forces that are acting on the object. If you pull on the handle of a drawer, the drawer will move in the direction you pulled it. If a ball is rolling down a driveway and hits a curb, the force of the curb will cause the ball to stop. Applying a pull or push to an object is an example of a contact force, where one object applies a force to another object through direct contact. There are other types of forces that can act on objects that do not involve objects touching. For example, the magnetic force produced by a magnet can make a paper clip move toward it or make another magnet move away from it without touching them. Another example is static electricity. Static electricity in a rubber balloon can cause a person's hair to stand up without the balloon actually touching any of his or her hair. Magnetic forces and electrical forces are therefore called non-contact forces because they can act on objects at a distance. Perhaps the most common non-contact force is gravity. Gravity is a force of attraction between two objects; the force due to gravity always works to bring objects closer together.

Any two objects, as long as they have some mass, will have a gravitational force of attraction between them. The force of gravity that exists between any two objects is influenced by the masses of those two objects. The mass of an object refers to the amount of matter that is contained by the object. Mass is also a measure of inertia, which is the resistance an object has to a change in its state of motion. All objects resist changes in their state of motion; however, the more massive an object, the more it will resist changes in its state of motion. The distance between any two objects will also influence the force of gravity that exists between them because the distance between any two objects can, and does, change. The exact relationship between these factors, however, was not well understood until 1687.

The first person to determine how mass and distance affect the strength of the gravitational force that exists between two objects was Isaac Newton. Newton described the relationship between these three factors in the book Philosophiae Naturalis Principia Mathematica (Newton 1687). The ability to describe the relationship between mass, distance, and the strength of a gravitational force was a major milestone in physics. It not only explained why objects fall toward the center of Earth (see Figure L6.1) but also explained why the planets move around the Sun, which was established by Copernicus in 1543 (see Figure L6.2). Before Newton put forth his revolutionary ideas about gravity, many people thought objects on Earth and objects in the sky moved because of different forces. Newton was the first person to suggest that the force of gravity is universal.

# Strength of Gravitational Force

*How Does the Gravitational Force That Exists Between Two Objects Relate to Their Masses and the Distance Between Them?*

In this investigation, you will have an opportunity to explore the relationship between mass, distance, and the strength of the gravitational force that exists between two objects in order to learn more about the behavior of gravity. This type of investigation can be difficult, however, because identifying the exact nature of the relationship that exists between several different factors is challenging. Take mass as an example. There are many potential ways that the strength of a gravitational force between two objects can be related to the mass of those two objects. The strength of the gravitational force between two objects may depend on the mass of the larger object or the mass of the smaller object. The strength of the gravitational force could also be related to the total mass of the two objects. In addition to mass, there are many different ways that the strength of a gravitational force between two objects can be related to the distance between the two objects. The strength of a gravitational force may increase as the distance between the two objects increases, or it may decrease as the distance between the two objects increases. It may also increase or decrease exponentially as the distance between the two objects changes. All of these different relationships are possible (along with many others). Your goal is to figure out the actual relationship.

## Your Task

Use what you know about forces, motion, patterns, and proportional relationships to design and carry out an investigation using a simulation to determine the relationship between mass, distance, and the strength of a gravitational force.

The guiding question of this investigation is, **How does the gravitational force that exists between two objects relate to their masses and the distance between them?**

## FIGURE L6.1

**Objects fall toward the center of Earth.**

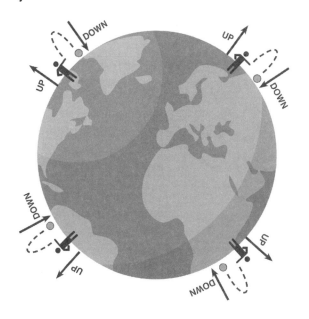

## FIGURE L6.2

**The heliocentric universe proposed by Copernicus in 1543**

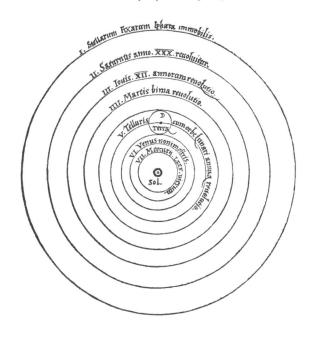

# LAB 6

## Materials

You will use an online simulation called *Gravity Force Lab* to conduct your investigation. You can access the simulation by going to the following website: *http://phet.colorado.edu/en/simulation/gravity-force-lab*.

## Safety Precautions

Follow all normal lab safety rules.

## Investigation Proposal Required?    ☐ Yes       ☐ No

## Getting Started

The *Gravity Force Lab* simulation (see screen shot in Figure L6.3) enables you to measure the amount of gravitational force that two objects exert on each other. You can adjust the mass of the two different objects in the simulation and the amount of distance between them. As you change mass and distance, you will be able to see the amount of force, in newtons, that each object exerts on the other one. To use this simulation, start by clicking on the "Show values" box in the lower-right corner of the window. This will allow you to see the amount of force exerted by the blue object (m1) on the red object (m2) and the amount of force exerted by the red object (m2) on the blue object (m1). You can change the masses of the blue and red objects by using the sliders at the bottom of the window. To change the distance between the red and blue objects, you can simply drag and drop each one to a different spot above the ruler. This simulation is useful because it allows you to measure the force of gravity under different conditions, and perhaps more important, it provides a way for you to design and carry out controlled experiments so you can focus on one factor at a time.

You will need to design and carry out at least three different experiments using the *Gravity Force Lab* simulation in order to determine the relationship between mass, distance, and gravitational force. You will need to conduct three different experiments because you will need to be able to answer three specific questions before you will be able to develop an answer to the guiding question:

1. How does changing the mass of the red object (m2) affect the amount of gravitational force?

2. How does changing the mass of the blue object (m1) affect the amount of gravitational force?

3. How does changing the distance between the two objects affect the amount of gravitational force?

National Science Teachers Association

# Strength of Gravitational Force

*How Does the Gravitational Force That Exists Between Two Objects Relate to Their Masses and the Distance Between Them?*

# FIGURE L6.3

**A screen shot of the *Gravity Force Lab* simulation**

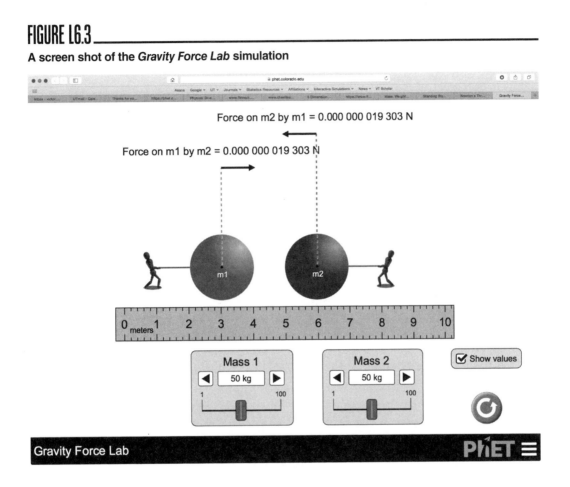

You will also need to determine what type of data you need to collect, how you will collect it, and how you will analyze the data for each experiment, because each experiment is slightly different.

To determine *what type of data you need to collect,* think about the following questions:

- What type of measurements will you need to record during each experiment?
- When will you need to make these measurements or observations?

To determine *how you will collect the data* using the simulation, think about the following questions:

- What will serve as your dependent variable for each experiment?
- What will serve as your independent variable for each experiment?
- How will you vary the independent variable during each experiment?
- What will you do to hold the other variables constant during each experiment?

- What types of comparisons will you need to make using the simulation?
- How many comparisons will you need to make to determine a trend or a relationship?
- How will you keep track of the data you collect and how will you organize it?

To determine *how you will analyze the data*, think about the following questions:

- What type of calculations will you need to make?
- What type of graph could you create to help make sense of your data?

Once you have carried out all your different experiments, your group will need to use your findings to develop an answer to the guiding question for this investigation. Your answer to the guiding question will need to be able to explain how the gravitational force that exists between two objects is related to the masses of the objects and distance between them. For your claim to be sufficient, your answer will need to be based on findings from all three of your experiments. You can then transform the data you collected during each experiment into evidence to support the validity of your overall explanation.

## Connections to Crosscutting Concepts, the Nature of Science, and the Nature of Scientific Inquiry

As you work through your investigation, be sure to think about

- the importance of looking for and understanding patterns in data,
- the importance of understanding proportional relationships in science,
- how scientific knowledge can change over time, and
- the culture of science and how it influences the work of scientists.

## Initial Argument

Once your group has finished collecting and analyzing your data, your group will need to develop an initial argument. Your initial argument needs to include a *claim*, *evidence* to support your claim, and a *justification* of the evidence. The claim is your group's answer to the guiding question. The evidence is an analysis and interpretation of your data. Finally, the justification of the evidence is why your group thinks the evidence matters. The justification of the evidence is important because scientists can use different kinds of evidence to support their claims. Your group will create your initial argument on a whiteboard. Your whiteboard should include all the information shown in Figure L6.4.

# Strength of Gravitational Force

*How Does the Gravitational Force That Exists Between Two Objects Relate to Their Masses and the Distance Between Them?*

## Argumentation Session

The argumentation session allows all of the groups to share their arguments. One member of each group will stay at the lab station to share that group's argument, while the other members of the group go to the other lab stations to listen to and critique the arguments developed by their classmates. This is similar to how scientists present their arguments to other scientists at conferences. If you are responsible for critiquing your classmates' arguments, your goal is to look for mistakes so these mistakes can be fixed and they can make their argument better. The argumentation session is also a good time to think about ways you can make your initial argument better. Scientists must share and critique arguments like this to develop new ideas.

## FIGURE L6.4

**Argument presentation on a whiteboard**

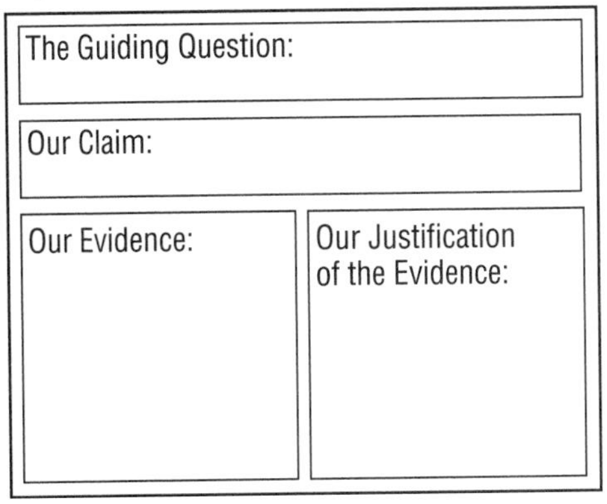

| The Guiding Question: | |
|---|---|
| Our Claim: | |
| Our Evidence: | Our Justification of the Evidence: |

To critique an argument, you might need more information than what is included on the whiteboard. You will therefore need to ask the presenter lots of questions. Here are some good questions to ask:

- How did you collect your data? Why did you use that method? Why did you collect those data?

- What did you do to make sure the data you collected are reliable? What did you do to decrease measurement error?

- How did your group analyze the data? Why did you decide to do it that way? Did you check your calculations?

- Is that the only way to interpret the results of your analysis? How do you know that your interpretation of your analysis is appropriate?

- Why did your group decide to present your evidence in that way?

- What other claims did your group discuss before you decided on that one? Why did your group abandon those alternative ideas?

- How confident are you that your claim is valid? What could you do to increase your confidence?

Once the argumentation session is complete, you will have a chance to meet with your group and revise your initial argument. Your group might need to gather more data or design a way to test one or more alternative claims as part of this process. Remember, your goal at this stage of the investigation is to develop the most acceptable and valid answer to the research question!

# LAB 6

## Report

Once you have completed your research, you will need to prepare an *investigation report* that consists of three sections. Each section should provide an answer to the following questions:

1. What question were you trying to answer and why?

2. What did you do to answer your question and why?

3. What is your argument?

Your report should answer these questions in two pages or less. This report must be typed, and any diagrams, figures, or tables should be embedded into the document. Be sure to write in a persuasive style; you are trying to convince others that your claim is acceptable and valid!

## References

Copernicus, N. 1543. De revolutionibus orbium coelestium [On the revolutions of heavenly spheres]. Johannes Petreius.

Newton, I. 1687. Philosophiae naturalis principia mathematica [Mathematical principles of natural philosophy].

**Strength of Gravitational Force**

*How Does the Gravitational Force That Exists Between Two Objects Relate to Their Masses and the Distance Between Them?*

*Checkout Questions*

# Lab 6. Strength of Gravitational Force
## How Does the Gravitational Force That Exists Between Two Objects Relate to Their Masses and the Distance Between Them?

1. The diagrams below show two objects and the distance between them.

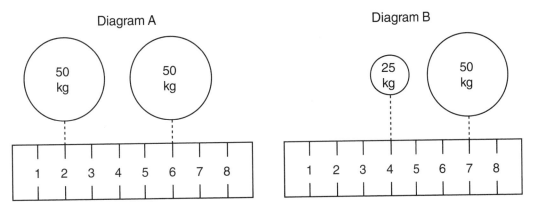

Which two objects have the greater gravitational attraction between them?

a. The objects in diagram A

b. The objects in diagram B

c. The gravitational attraction between the objects is the same in diagrams A and B

d. Unsure

How do you know?

2. The diagrams below show two objects and the distance between them.

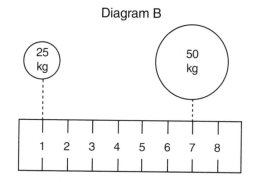

Diagram A

Diagram B

Which two objects have the greater gravitational attraction between them?

a. The objects in diagram A

b. The objects in diagram B

c. The gravitational attraction between the objects is the same in diagrams A and B.

d. Unsure

How do you know?

3. Once a scientist or a team of scientists develop a scientific law, it will never change.

a. I agree with this statement.

b. I disagree with this statement.

Explain your answer, using an example from your investigation about gravitational force.

4. Cultural values and expectations determine who gets to do science, what scientists choose to investigate, how investigations are conducted, and how research findings are interpreted.

    a. I agree with this statement.
    b. I disagree with this statement.

    Explain your answer, using an example from your investigation about gravitational force.

5. Scientists often need to look for and understand the underlying cause of patterns in data. Explain why it is important to be able to identify and understand patterns in data in science, using an example from your investigation about gravitational force.

6. Scientists often look for proportional relationships between two or more quantities in science. Explain what proportional relationships are and why they are important in science, using an example from your investigation about gravitational force.

*Lab Handout*

# Lab 7. Mass and Free Fall

## How Does Mass Affect the Amount of Time It Takes for an Object to Fall to the Ground?

### Introduction

The motion of an object is the result of all the different forces that are acting on the object. If you push a toy car across the floor, it moves in the direction you pushed it. If the car then hits a wall, the force of the wall causes the car to stop. Applying a push or a pull to an object is an example of a contact force, where one object applies a force to another object through direct contact. There are other types of forces that can act on objects that do not involve objects touching. For example, a strong magnet can pull on a paper clip and make it move without ever actually touching the paper clip. Another example is static electricity. Static electricity in a rubber balloon can cause a person's hair to stand up without the balloon actually touching any of his or her hair. Magnetic forces and electrical forces are therefore called non-contact forces. Perhaps the most common non-contact force is gravity. Gravity is a force of attraction between two objects; the force due to gravity always works to bring objects closer together.

Any two objects that have mass will also have a gravitational force of attraction between them. Consider the Sun, Earth, and Moon as examples. Earth and the Moon are very large and have a lot of mass; the force of gravity between Earth and the Moon is strong enough to keep the Moon orbiting Earth even though they are very far apart. Similarly, the force of gravity between the Sun and Earth is strong enough to keep Earth in orbit around the Sun, despite Earth and the Sun being millions of miles apart. The force of gravity between two objects depends on the amount of mass of each object and how far apart they are. Objects that are more massive produce a greater gravitational force. The force of gravity between two objects also weakens as the distance between the two objects increases. So even though Earth and the Sun are very far apart from each other (which means less gravity), the fact that they are both very massive (which means more gravity) results in a gravitational force that is strong enough to keep Earth in orbit.

The gravitational force that acts between two objects, as noted earlier, can cause one of those objects to move. The skydivers in Figure L7.1, for example, are moving toward the center of Earth because of the force of gravity. Scientists describe the motion of an object by describing its speed, velocity, and acceleration. Speed is the distance an object travels in a specific amount of time. Velocity is the speed of an object in a given direction. Acceleration is the change in velocity divided by time. The amount of force required to change

## FIGURE L7.1

**These skydivers are falling toward Earth because of gravity.**

the motion of an object depends on the mass of that object. Therefore, as the mass of an object increases, so does the amount of force that is needed to change its motion.

In this investigation you will have an opportunity to explore the relationship between mass and the time it takes for an object to fall to the ground. Many people think that heavier objects fall to the ground faster than lighter ones because gravity will pull on heavier objects with more force, so the heavier object will accelerate faster. Other people, however, think that heavier objects have more inertia (the tendency of an object to stay still if it is still or keep moving if it is currently moving) so heavier objects will be less responsive to the force of gravity and take longer to accelerate. Unfortunately, it is challenging to determine which of these two explanations is the most valid because objects encounter air resistance as they fall. Air resistance is the result of an object moving through a layer of air and colliding with air molecules. The more air molecules that an object collides with, the greater the air resistance force. Air resistance is therefore dependent on the speed of the falling object and the surface area of the falling object. Since massive objects are often larger than less massive ones (consider a bowling ball and a marble as an example), it is often difficult to design a fair test of these two explanations. To determine the relationship between mass and the time it takes an object to fall to the ground, you will therefore need to design an experiment that will allow you to control for the influence of air resistance.

## Your Task

Use what you know about forces and motion, patterns, and rates of change to design and carry out an experiment to determine the relationship between mass and the time it takes an object to fall to the ground.

The guiding question of this investigation is, **How does mass affect the amount of time it takes for an object to fall to the ground?**

## Materials

You may use any of the following materials during your investigation:

- Beanbag A
- Beanbag B
- Beanbag C
- Meterstick

- Stopwatch
- Electronic or triple beam balance
- Masking tape
- Safety glasses or goggles

## Safety Precautions

Follow all normal lab safety rules. In addition, take the following safety precautions:

1. Wear sanitized safety glasses or goggles during lab setup, hands-on activity, and takedown.

2. Do not throw the beanbags.

# LAB 7

3. Do not stand on tables or chairs.

4. Wash hands with soap and water after completing the lab activity.

**Investigation Proposal Required?**   ☐ Yes      ☐ No

## Getting Started

To answer the guiding question, you will need to design and conduct an experiment as part of your investigation. To accomplish this task, you must determine what type of data you need to collect, how you will collect it, and how you will analyze it.

To determine *what type of data you need to collect,* think about the following questions:

- What will serve as your independent variable?
- What will serve as your dependent variable?
- What measurements will you need to determine the rate of a falling object?

To determine *how you will collect your data,* think about the following questions:

- What variables will need to be controlled and how will you control them?
- How many tests will you need to run to have reliable data (to make sure it is consistent)?
- How will you make sure that your data are of high quality (i.e., how will you reduce error)?
- How will you keep track of the data you collect, and how will you organize it?

To determine *how you will analyze your data,* think about the following questions:

- How will you calculate the rate of a falling object?
- What type of calculations will you need to make to take into account multiple trials?
- What types of graphs or tables could you create to help make sense of your data?

## Connections to Crosscutting Concepts, the Nature of Science, and the Nature of Scientific Inquiry

As you work through your investigation, be sure to think about

- the importance of looking for and understanding patterns in data,
- the importance of understanding proportional relationships in science,
- the difference between laws and theories in science, and
- the difference between data and evidence in science.

National Science Teachers Association

## Initial Argument

Once your group has finished collecting and analyzing your data, your group will need to develop an initial argument. Your initial argument needs to include a *claim, evidence* to support your claim, and a *justification* of the evidence. The claim is your group's answer to the guiding question. The evidence is an analysis and interpretation of your data. Finally, the justification of the evidence is why your group thinks the evidence matters. The justification of the evidence is important because scientists can use different kinds of evidence to support their claims. Your group will create your initial argument on a whiteboard. Your whiteboard should include all the information shown in Figure L7.2.

## FIGURE L7.2 _____

**Argument presentation on a whiteboard**

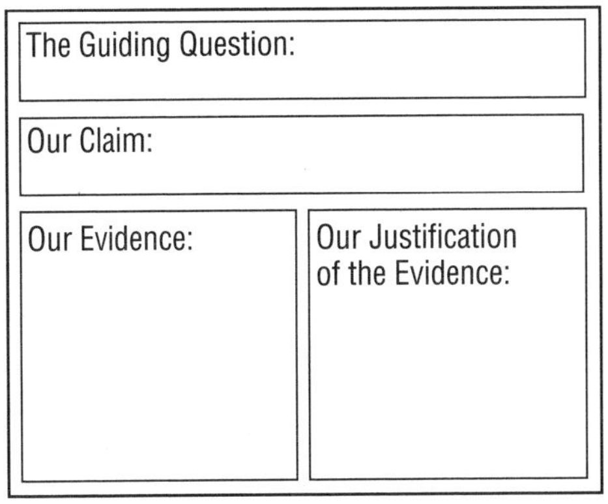

| The Guiding Question: |  |
|---|---|
| Our Claim: |  |
| Our Evidence: | Our Justification of the Evidence: |

## Argumentation Session

The argumentation session allows all of the groups to share their arguments. One member of each group will stay at the lab station to share that group's argument, while the other members of the group go to the other lab stations to listen to and critique the arguments developed by their classmates. This is similar to how scientists present their arguments to other scientists at conferences. If you are responsible for critiquing your classmates' arguments, your goal is to look for mistakes so these mistakes can be fixed and they can make their argument better. The argumentation session is also a good time to think about ways you can make your initial argument better. Scientists must share and critique arguments like this to develop new ideas.

To critique an argument, you might need more information than what is included on the whiteboard. You will therefore need to ask the presenter lots of questions. Here are some good questions to ask:

- How did you collect your data? Why did you use that method? Why did you collect those data?
- What did you do to make sure the data you collected are reliable? What did you do to decrease measurement error?
- How did your group analyze the data? Why did you decide to do it that way? Did you check your calculations?
- Is that the only way to interpret the results of your analysis? How do you know that your interpretation of your analysis is appropriate?
- Why did your group decide to present your evidence in that way?
- What other claims did your group discuss before you decided on that one? Why did your group abandon those alternative ideas?

- How confident are you that your claim is valid? What could you do to increase your confidence?

Once the argumentation session is complete, you will have a chance to meet with your group and revise your initial argument. Your group might need to gather more data or design a way to test one or more alternative claims as part of this process. Remember, your goal at this stage of the investigation is to develop the most acceptable and valid answer to the research question!

## Report

Once you have completed your research, you will need to prepare an *investigation report* that consists of three sections. Each section should provide an answer to the following questions:

1. What question were you trying to answer and why?

2. What did you do to answer your question and why?

3. What is your argument?

Your report should answer these questions in two pages or less. This report must be typed, and any diagrams, figures, or tables should be embedded into the document. Be sure to write in a persuasive style; you are trying to convince others that your claim is acceptable and valid!

## Checkout Questions

# Lab 7. Mass and Free Fall

## How Does Mass Affect the Amount of Time It Takes for an Object to Fall to the Ground?

1. A group of students is investigating the relationship between mass and the time it takes for an object to reach the ground. They have six different cubes. Each cube is the same size but a different mass. They label the cubes in order of their relative mass. Cube A is the heaviest and cube F is the lightest. They then drop each cube from a height of 5 meters and time how long it takes for each cube to hit the ground. Use Tables 1–3 to answer the question below.

| Table 1 | | Table 2 | | Table 3 | |
|---|---|---|---|---|---|
| **Cube** | **Time (seconds)** | **Cube** | **Time (seconds)** | **Cube** | **Time (seconds)** |
| A | 1.3 | A | 1.0 | A | 0.8 |
| B | 1.2 | B | 1.1 | B | 0.9 |
| C | 1.1 | C | 1.0 | C | 1.0 |
| D | 1.0 | D | 0.9 | D | 1.1 |
| E | 0.9 | E | 1.0 | E | 1.2 |
| F | 0.8 | F | 1.0 | F | 1.3 |

Which table do you think best represents the data that the students would have collected?

a. Table 1

b. Table 2

c. Table 3

d. Unsure

How do you know?

2. "The force of gravitational attraction is directly dependent on the masses of both objects and inversely proportional to the square of the distance that separates their centers" is an example of a scientific theory.

   a.  I agree with this statement.

   b.  I disagree with this statement.

Explain your answer, using an example from your investigation about mass and free fall.

3. "It took the cube 1.1 seconds to reach the ground" is an example of evidence.

   a.  I agree with this statement.

   b.  I disagree with this statement.

Explain your answer, using an example from your investigation about mass and free fall.

4. Scientists often need to look for and understand the underlying cause of patterns in data. Explain why it important to be able to identify and understand patterns in data in science, using an example from your investigation about mass and free fall.

5. Scientists often look for proportional relationships between quantities in science. Explain what proportional relationships are and why they are important in science, using an example from your investigation about mass and free fall.

*Lab Handout*

# Lab 8. Force and Motion
## How Do Changes in Pulling Force Affect the Motion of an Object?

### Introduction

A force can be described simply as a push or pull that acts on an object. For example, when you push or pull on a doorknob, you are applying a force that moves the door. In addition to a push or a pull, forces can be described as contact or non-contact. Pushing a box across the floor is an example of a contact force; the force to move the box is being applied by your hands, which are in contact with the box. However, non-contact forces can act on objects without having to actually touch the object. For example, a magnet can push or pull another magnet without the two ever touching each other. Similarly, gravity is a non-contact force that pulls objects closer together, such as when something falls toward Earth.

When you apply a force to an object, that object often will move. Sometimes, however, when you apply a force to an object it doesn't move. It is relatively easy to apply enough force to slide a box across the floor, but it is much more difficult to push a car down the road. The motion of an object is determined by the strength of the force applied to move it, the weight of the object, and any other forces that might be acting to move the object in a different direction. Consider a game of tug-of-war (see Figure L8.1): if both people pull with equal strength, then the rope doesn't move, but if one person pulls harder, the rope moves in that direction.

# FIGURE L8.1 _____

**In a game of tug-of-war, the overall movement of the rope is based on the strength of the pull in both directions.**

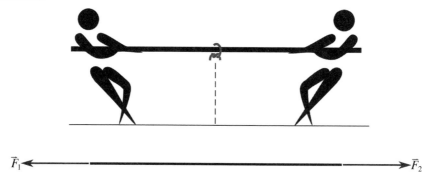

Isaac Newton (1642–1727) was a physicist who studied the motion of objects. He is perhaps most well known for the laws of motion he developed after extensive observation of the planets in our solar system. Newton described that (1) a stationary object will remain stationary unless an external force acts on it, (2) the change in an object's motion is proportional to the force acting on it, and (3) every force has an equal and opposite force. The motion of an object is the result of all the different forces that are pushing or pulling on that object. When all the forces are acting in the same direction, the object will move in that direction. If all the forces acting on an object are balanced, then the object either will not move or will move with a constant speed. When there are forces acting in different directions on the same object but they are not the same strength, then the forces are unbalanced; the object will move, but how it moves (e.g., fast, slow, constant speed, speeds up, or slows down) depends on the relationship of all the forces acting on the object.

## Your Task

Use what you know about forces, systems, and stability and change to design and carry out an investigation that will allow you to predict how different pulling forces (hanging weights) influence the motion of a cart (e.g., does it speed up, slow down, or travel at a constant speed).

The guiding question of this investigation is, **How do changes in pulling force affect the motion of an object?**

## Materials

You may use any of the following materials during your investigation:

- Pull cart
- Pull car track or flat table
- Pulley
- Pulley clamp
- String

- Hanging weights
- Meterstick
- Electronic or triple beam balance
- Motion sensor with interface
- Safety glasses or goggles

## Safety Precautions

Follow all normal lab safety rules. In addition, take the following safety precautions:

1. Wear sanitized safety glasses or goggles during lab setup, hands-on activity, and takedown.

2. Use caution when working with the moving pulleys, cart, and weights.

3. Keep your fingers and toes out of the way of the moving objects.

4. Wash hands with soap and water after completing the lab activity.

## Investigation Proposal Required?     ☐ Yes          ☐ No

# LAB 8

## FIGURE L8.2

**Setting up the motion sensor, cart, and pulley**

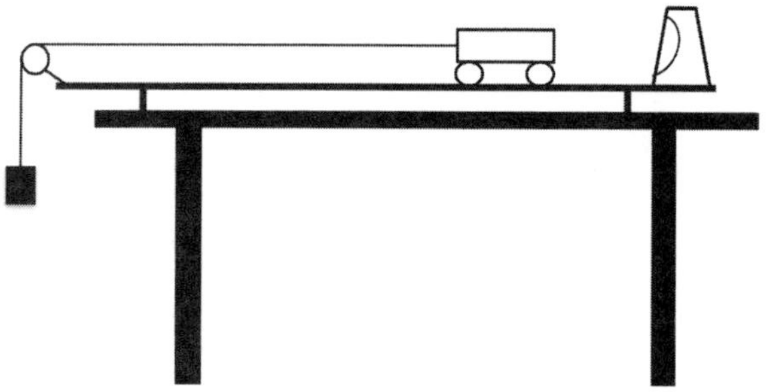

## Getting Started

To answer the guiding question, you will need to plan an investigation to measure the motion of a cart as it is pulled across the tabletop. Figure L8.2 shows how you can set up the cart and motion sensor to collect your data; however, to accomplish this task, you must determine what type of data you need to collect, how you will collect it, and how you will analyze it.

To determine *what type of data you need to collect,* think about the following questions:

- What information do you need to describe the motion of the cart?
- What information or measurements do you need to calculate the speed of the cart?

To determine *how you will collect your data,* think about the following questions:

- What equipment will you need to collect the data you need?
- How will you make sure that your data are of high quality (i.e., how will you reduce error)?
- How will you keep track of the data you collect?
- How will you organize your data?

To determine *how you will analyze your data,* think about the following questions:

- What type of calculations will you need to make?
- What type of table or graph could you create to help make sense of your data?
- How will you determine the effect of different pulling forces on the cart's motion?

## Connections to Crosscutting Concepts, the Nature of Science, and the Nature of Scientific Inquiry

As you work through your investigation, be sure to think about

- how scientists often need to understand and define systems under study and that making a model is a tool for developing a better understanding of natural phenomena in science,
- the importance of understanding what makes a system stable or unstable and what controls rates of change within a system,
- how scientists use different types of methods to answer different types of questions, and
- the nature and role of experiments within science.

## Initial Argument

Once your group has finished collecting and analyzing your data, your group will need to develop an initial argument. Your initial argument needs to include a *claim, evidence* to support your claim, and a *justification* of the evidence. The claim is your group's answer to the guiding question. The evidence is an analysis and interpretation of your data. Finally, the justification of the evidence is why your group thinks the evidence matters. The justification of the evidence is important because scientists can use different kinds of evidence to support their claims. Your group will create your initial argument on a whiteboard. Your whiteboard should include all the information shown in Figure L8.3.

## Argumentation Session

The argumentation session allows all of the groups to share their arguments. One member of each group will stay at the lab station to share that group's argument, while the other members of the group go to the other lab stations to listen to and critique the arguments developed by their classmates. This is similar to how scientists present their arguments to other scientists at conferences. If you are responsible for critiquing your classmates' arguments, your goal is to look for mistakes so these mistakes can be fixed and they can make their argument better. The argumentation session is also a good time to think about ways you can make your initial argument better. Scientists must share and critique arguments like this to develop new ideas.

## FIGURE L8.3

**Argument presentation on a whiteboard**

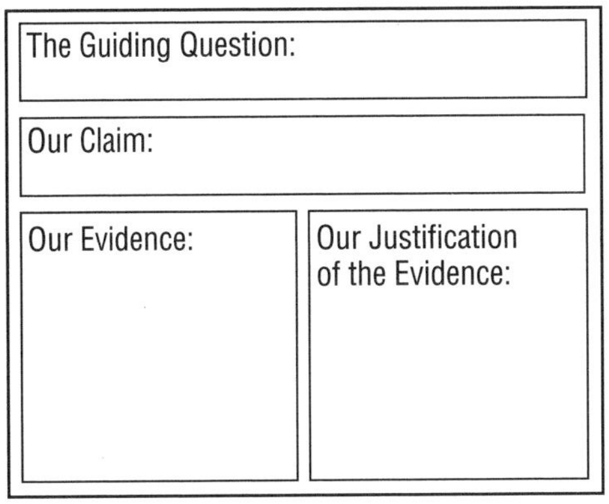

| The Guiding Question: | |
|---|---|
| Our Claim: | |
| Our Evidence: | Our Justification of the Evidence: |

To critique an argument, you might need more information than what is included on the whiteboard. You will therefore need to ask the presenter lots of questions. Here are some good questions to ask:

- How did you collect your data? Why did you use that method? Why did you collect those data?

- What did you do to make sure the data you collected are reliable? What did you do to decrease measurement error?

- How did your group analyze the data? Why did you decide to do it that way? Did you check your calculations?

- Is that the only way to interpret the results of your analysis? How do you know that your interpretation of your analysis is appropriate?

- Why did your group decide to present your evidence in that way?

- What other claims did your group discuss before you decided on that one? Why did your group abandon those alternative ideas?

- How confident are you that your claim is valid? What could you do to increase your confidence?

Once the argumentation session is complete, you will have a chance to meet with your group and revise your initial argument. Your group might need to gather more data or design a way to test one or more alternative claims as part of this process. Remember, your goal at this stage of the investigation is to develop the most acceptable and valid answer to the research question!

## Report

Once you have completed your research, you will need to prepare an *investigation report* that consists of three sections. Each section should provide an answer to the following questions:

1. What question were you trying to answer and why?

2. What did you do to answer your question and why?

3. What is your argument?

Your report should answer these questions in two pages or less. This report must be typed, and any diagrams, figures, or tables should be embedded into the document. Be sure to write in a persuasive style; you are trying to convince others that your claim is acceptable and valid!

## *Checkout Questions*

# Lab 8. Force and Motion
## How Do Changes in Pulling Force Affect the Motion of an Object?

1. Describe a general rule for predicting the motion of an object that is being pushed or pulled by unbalanced forces.

2. Below is a position versus time graph for a car accelerating away from a stoplight.

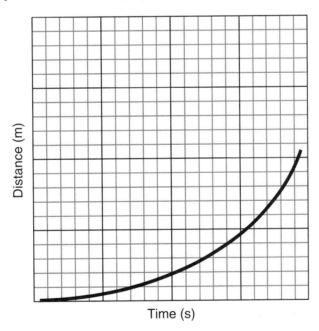

Time (s)

Draw a velocity versus time graph for the same car.

Explain your answer. Why did you draw your graph that way?

3. Experiments are the best way to get answers during a scientific investigation.

    a. I agree with this statement.

    b. I disagree with this statement.

Explain your answer, using an example from your investigation about force and motion.

4. The scientific method guides scientists when they do their work.

    a. I agree with this statement.

    b. I disagree with this statement.

Explain your answer, using an example from your investigation about force and motion.

5. Scientists sometimes study systems that are very large or very small, and sometimes scientists study systems that have lots of components. It is useful to make models of complex systems to better understand what is going on. Explain how models can be useful for understanding systems, using an example from your investigation about force and motion.

6. It is often important to understand the relationships between components of a system. Explain why it is important to identify factors that cause a system to become unstable, using an example from your investigation about force and motion.

# LAB 9

## *Lab Handout*

## Lab 9. Mass and Motion
### How Do Changes in the Mass of an Object Affect Its Motion?

### Introduction

The motion of an object depends on all the different forces that are acting on the object, how strong those different forces are, and how much mass the object has. Changing the number of forces acting on an object, the direction of those forces, or the strength of the forces will have an effect on the object's motion. But what happens if you apply all the same forces to two different objects, one with a small mass and one with a larger mass?

The amount of mass of an object is a measure of the amount of matter in that object. Objects with more mass have more weight (a measure of the force of gravity) because there is a stronger attraction between the object and Earth due to gravity. Motorcycles are light-weight vehicles, about 250 kg, with strong engines that help them travel at high speeds (see Figure L9.1). How might the speed of a motorcycle change if it is carrying one rider or two riders? In this example the strength of the engine doesn't change, but the total mass of the motorcycle and riders increases. It is important for scientists and engineers to understand the relationship between the forces applied to an object and the mass of that object so that they can predict the object's motion.

## FIGURE L9.1 _____

**Motorcycles are lightweight, but their engines generate a lot of force.**

### Your Task

Use what you know about forces, stability and change, and patterns to design and conduct an investigation that will allow you to describe the motion of a cart (e.g., does it speed up, slow down, or travel at a constant speed) and how its motion is affected by changing the mass of the cart while keeping the pulling force the same.

The guiding question of this investigation is, **How do changes in the mass of an object affect its motion?**

**90**

National Science Teachers Association

## Materials

You may use any of the following materials during your investigation:

- Pull cart
- Pull car track or flat table
- Pulley
- Pulley clamp
- Cart masses
- String

- Hanging weights
- Meterstick
- Electronic or triple beam balance
- Motion sensor with interface
- Safety glasses or goggles

## Safety Precautions

Follow all normal lab safety rules. In addition, take the following safety precautions:

1. Wear sanitized safety glasses or goggles during lab setup, hands-on activity, and takedown.

2. Use caution when working with the moving pulleys, cart, masses, and weights.

3. Keep your fingers and toes out of the way of the moving objects.

4. Wash hands with soap and water after completing the lab activity.

## Investigation Proposal Required?     ☐ Yes          ☐ No

## Getting Started

To answer the guiding question, you will need to plan an investigation to measure the motion of a cart as it is pulled across the tabletop. Figure L9.2 shows how you can set up the cart and motion sensor to collect your data; however, to accomplish this task, you must determine what type of data you need to collect, how you will collect it, and how you will analyze it.

To determine *what type of data you need to collect,* think about the following questions:

- What information do you need to describe the motion of the cart?

- What information or measurements do you need to calculate the speed of the cart?

To determine *how you will collect your data,* think about the following questions:

- What equipment will you need to collect the data you need?

## FIGURE L9.2 _____

**Setting up the motion sensor, cart, pulley, and masses**

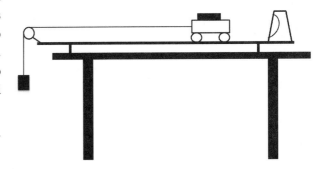

- How will you make sure that your data are of high quality (i.e., how will you reduce error)?
- How will you keep track of the data you collect?
- How will you organize your data?

To determine *how you will analyze your data,* think about the following questions:

- What type of calculations will you need to make?
- What type of table or graph could you create to help make sense of your data?
- How will you determine the effect of different pulling forces on the cart's motion?

## Connections to Crosscutting Concepts, the Nature of Science, and the Nature of Scientific Inquiry

As you work through your investigation, be sure to think about

- how scientists often look for patterns in nature and attempt to understand the underlying cause of these patterns,
- the importance of understanding what makes a system stable or unstable and what controls rates of change within a system,
- the differences between observations and inferences in science, and
- the difference between data and evidence in science.

## Initial Argument

Once your group has finished collecting and analyzing your data, your group will need to develop an initial argument. Your initial argument needs to include a *claim, evidence* to support your claim, and a *justification* of the evidence. The claim is your group's answer to the guiding question. The evidence is an analysis and interpretation of your data. Finally, the justification of the evidence is why your group thinks the evidence matters. The justification of the evidence is important because scientists can use different kinds of evidence to support their claims. Your group will create your initial argument on a whiteboard. Your whiteboard should include all the information shown in Figure L9.3.

## FIGURE L9.3

**Argument presentation on a whiteboard**

| The Guiding Question: | |
|---|---|
| Our Claim: | |
| Our Evidence: | Our Justification of the Evidence: |

## Argumentation Session

The argumentation session allows all of the groups to share their arguments. One member of each group will stay at the lab station to share that group's argument, while the other members of the group go to the other lab stations to listen to and critique the arguments developed by their classmates. This is similar to how scientists present their arguments to other scientists at conferences. If you are responsible for critiquing your classmates' arguments, your goal is to look for mistakes so these mistakes can be fixed and they can make their argument better. The argumentation session is also a good time to think about ways you can make your initial argument better. Scientists must share and critique arguments like this to develop new ideas.

To critique an argument, you might need more information than what is included on the whiteboard. You will therefore need to ask the presenter lots of questions. Here are some good questions to ask:

- How did you collect your data? Why did you use that method? Why did you collect those data?

- What did you do to make sure the data you collected are reliable? What did you do to decrease measurement error?

- How did your group analyze the data? Why did you decide to do it that way? Did you check your calculations?

- Is that the only way to interpret the results of your analysis? How do you know that your interpretation of your analysis is appropriate?

- Why did your group decide to present your evidence in that way?

- What other claims did your group discuss before you decided on that one? Why did your group abandon those alternative ideas?

- How confident are you that your claim is valid? What could you do to increase your confidence?

Once the argumentation session is complete, you will have a chance to meet with your group and revise your initial argument. Your group might need to gather more data or design a way to test one or more alternative claims as part of this process. Remember, your goal at this stage of the investigation is to develop the most acceptable and valid answer to the research question!

## Report

Once you have completed your research, you will need to prepare an *investigation report* that consists of three sections. Each section should provide an answer to the following questions:

1. What question were you trying to answer and why?

2. What did you do to answer your question and why?

3. What is your argument?

Your report should answer these questions in two pages or less. This report must be typed, and any diagrams, figures, or tables should be embedded into the document. Be sure to write in a persuasive style; you are trying to convince others that your claim is acceptable and valid!

## *Checkout Questions*

# Lab 9. Mass and Motion
## How Do Changes in the Mass of an Object Affect Its Motion?

1. Describe a general relationship between the acceleration of two objects that are being pushed by the same-strength force, but one object is twice as heavy as the other.

2. Ashley is a race car driver and she wants her car to go faster. She is trying to decide between two plans to increase the acceleration of her car. Her car has a mass of 900 kg, and when she races it can accelerate at 20 m/s². Ashley would like her car to accelerate at 25 m/s². Her first plan is to make her car lighter by using some new materials; if she does that her car will have a mass of 675 kg. Her second plan is to get a stronger engine; the new engine would be 25% stronger than her current engine. But the new engine will make the car weigh 1,100 kg. Ashley only has enough money for one option. Which would you recommend?

   Explain your answer. Why did you make that recommendation?

3. In science, observations and inferences are the same thing.

   a. I agree with this statement.

   b. I disagree with this statement.

   Explain your answer, using an example from your investigation about mass and the motion of objects.

4. When discussing the results of an investigation, there is no need to differentiate between data and evidence—they are really the same thing.

   a. I agree with this statement.

   b. I disagree with this statement.

   Explain your answer, using an example from your investigation about mass and the motion of objects.

5. Identifying patterns in nature is important to the work of many scientists. Explain how understanding patterns and their causes is helpful to scientists. Use an example from your investigation about mass and the motion of objects to help in your explanation.

6. It is important for scientist to make predictions about natural systems. Use an example from your investigation about mass and the motion of objects to explain why it is important to identify factors that cause changes in a system or cause the system to become unstable.

***Lab Handout***

# Lab 10. Magnetic Force
## How Is the Strength of an Electromagnet Affected by the Number of Turns of Wire in a Coil?

### Introduction

Magnets and magnetic fields are useful for many applications. For example, small permanent magnets and electromagnets are used in speakers that are found in cell phones or headphones used to listen to music. In a speaker, the changes in the magnetic field of the electromagnet cause parts of the speaker to vibrate, which produces the sounds we hear when we listen to music. The electromagnets in headphone speakers are small and fairly weak, but other electromagnets can be much larger and stronger, such as those used in junkyards to pick up and move old cars. Electromagnets are also used in the medical field in devices such as MRI (magnetic resonance imaging) machines. The powerful electromagnets in MRI machines influence the atoms in our bodies and allow doctors to create images that are useful in diagnosing injuries.

Permanent magnets, such as refrigerator magnets or those made from combinations of metals such as iron (Fe), nickel (Ni), or neodymium (Nd), always demonstrate magnetic properties. Permanent magnets are surrounded by a magnetic field. This magnetic field can influence other magnets or some materials (like some metals) and cause the objects to be pulled toward the magnet or pushed away from the magnet.

Magnetic fields can also be created when electricity passes through a wire. The electric current (moving electrical charges) in the wire creates a magnetic field surrounding the wire (see Figure L10.1). The magnetic field surrounding the wire is usually weak, but it can still have an effect on other magnets or materials. Coiling the wire will help concentrate the magnetic field on the inside of the coil (see Figure L10.2).

Turning a coil of wire into an electromagnet is as simple as wrapping the coil of wire around a piece of metal, such as an iron nail (see Figure L10.3, p. 99). When a wire is coiled around the nail, the magnetic field from the wire that is concentrated inside the coils magnetizes the iron nail and produces the electromagnet. Individual iron atoms can act like very small magnets, but inside a nail, the iron atoms point in random directions; therefore, the nail on its own does not act like a magnet. But when the iron atoms inside the nail are influenced by the magnetic field from the coil of wire, they change their alignment and point in similar directions. Only iron atoms inside the coil of wire will change their alignment, and the more atoms that point in the same direction, the greater the magnetic strength of the nail. The nail will only act like a magnet when the electric current is flowing through the wire; when the electric current stops, the iron atoms return to their original and random alignment and no longer act like a magnet.

## FIGURE L10.1 _____

**Magnetic field surrounding a wire**

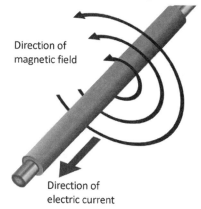

Direction of magnetic field

Direction of electric current

## FIGURE L10.2 _____

**Concentrated magnetic field in a coil of wire**

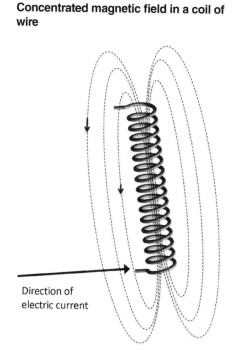

Direction of electric current

### Your Task

Build an electromagnet using a battery, some wire, and a nail. Then, use what you know about magnetic fields, tracking energy in a system, and structure and function to design and carry out an investigation that will allow you to determine how the number of turns of wire wrapped around the nail affects the strength of the electromagnet.

The guiding question of this investigation is, **How is the strength of an electromagnet affected by the number of turns of wire in a coil?**

### Materials

You may use any of the following materials during your investigation:

- Size D battery
- Battery holder
- Copper wire
- Iron nail
- Paper clips

- Gauss meter (optional)
- Electronic or triple beam balance
- Ruler
- Safety glasses or goggles

### Safety Precautions

Follow all normal lab safety rules. In addition, take the following safety precautions:

1. Wear sanitized safety glasses or goggles during lab setup, hands-on activity, and takedown.

2. Given that the electromagnet setup will generate heat, disconnect the wire from the battery when not actively collecting data, to prevent skin burn.

3. Use caution when wrapping the copper wire around the nail. It is possible that the loose end will swing around during the wrapping process and present a hazard.

4. Use caution in handling the wire end or nails. They are sharp and can cut or puncture skin.

5. Wash hands with soap and water after completing the lab activity.

## Investigation Proposal Required?  ☐ Yes  ☐ No

## Getting Started

To answer the guiding question, you will need to design and conduct an investigation to measure the strength of your electromagnet. To accomplish this task, you must determine what type of data you need to collect, how you will collect it, and how you will analyze it. Figure L10.3 shows how to construct a simple electromagnet from a battery, a copper wire, and a nail.

To determine *what type of data you need to collect,* think about the following questions:

- How will you determine the strength of the electromagnet?
- What information or measurements will you need to record?
- What parts of the electromagnet will you change and what parts will you keep consistent?

## FIGURE L10.3 _____

**Electromagnet made from a D-cell battery, a copper wire, and an iron nail**

To determine *how you will collect your data,* think about the following questions:

- What equipment will you need to collect the data you need?
- How will you make sure that your data are of high quality (i.e., how will you reduce error)?
- Are there different ways you can measure the amount of coils you used?
- How will you keep track of the data you collect?

# LAB 10

- How will you organize your data?

To determine *how you will analyze your data,* think about the following questions:

- How will you determine if the number of coils affects the strength of the electromagnet?
- What type of table or graph could you create to help make sense of your data?

## Connections to Crosscutting Concepts, the Nature of Science, and the Nature of Scientific Inquiry

As you work through your investigation, be sure to think about

- why it is important to track how energy and matter move into, out of, and within systems;
- how the structure or shape of something can influence how it functions and places limits on what it can and cannot do;
- the difference between theories and laws in science; and
- the difference between data and evidence in science.

## Initial Argument

Once your group has finished collecting and analyzing your data, your group will need to develop an initial argument. Your initial argument needs to include a *claim, evidence* to support your claim, and a *justification* of the evidence. The claim is your group's answer to the guiding question. The evidence is an analysis and interpretation of your data. Finally, the justification of the evidence is why your group thinks the evidence matters. The justification of the evidence is important because scientists can use different kinds of evidence to support their claims. Your group will create your initial argument on a whiteboard. Your whiteboard should include all the information shown in Figure L10.4.

## FIGURE L10.4 _____
**Argument presentation on a whiteboard**

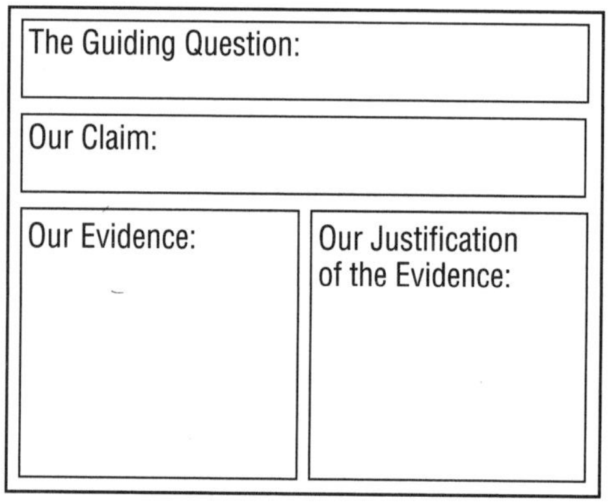

| The Guiding Question: | |
|---|---|
| Our Claim: | |
| Our Evidence: | Our Justification of the Evidence: |

## Argumentation Session

The argumentation session allows all of the groups to share their arguments. One member of each group will stay at the lab station to share that group's argument, while the other members of the group go to the other lab stations to listen to and critique the arguments developed by their classmates. This is similar to how scientists present their arguments to other scientists at conferences. If you are responsible for critiquing your classmates' arguments, your goal is to look for mistakes so these mistakes can be fixed and they

National Science Teachers Association

can make their argument better. The argumentation session is also a good time to think about ways you can make your initial argument better. Scientists must share and critique arguments like this to develop new ideas.

To critique an argument, you might need more information than what is included on the whiteboard. You will therefore need to ask the presenter lots of questions. Here are some good questions to ask:

- How did you collect your data? Why did you use that method? Why did you collect those data?

- What did you do to make sure the data you collected are reliable? What did you do to decrease measurement error?

- How did your group analyze the data? Why did you decide to do it that way? Did you check your calculations?

- Is that the only way to interpret the results of your analysis? How do you know that your interpretation of your analysis is appropriate?

- Why did your group decide to present your evidence in that way?

- What other claims did your group discuss before you decided on that one? Why did your group abandon those alternative ideas?

- How confident are you that your claim is valid? What could you do to increase your confidence?

Once the argumentation session is complete, you will have a chance to meet with your group and revise your initial argument. Your group might need to gather more data or design a way to test one or more alternative claims as part of this process. Remember, your goal at this stage of the investigation is to develop the most acceptable and valid answer to the research question!

## Report

Once you have completed your research, you will need to prepare an *investigation report* that consists of three sections. Each section should provide an answer to the following questions:

1. What question were you trying to answer and why?

2. What did you do to answer your question and why?

3. What is your argument?

Your report should answer these questions in two pages or less. This report must be typed, and any diagrams, figures, or tables should be embedded into the document. Be sure to write in a persuasive style; you are trying to convince others that your claim is acceptable and valid!

## Checkout Questions

# Lab 10. Magnetic Force

## How Is the Strength of an Electromagnet Affected by the Number of Turns of Wire in a Coil?

1. Malik and Jason are both making electromagnets. Malik wants to use two batteries to make his electromagnet the strongest. Jason plans on using only one battery but wrapping his wire twice as much as Malik. Use what you know about electromagnets to explain why both students have good strategies for making strong electromagnets.

2. Below are two samples of different materials. The samples show the general alignment of atoms within that material. Each individual atom acts like a miniature magnet, with the light gray side representing the south end of the magnet and the dark gray side representing the north end of the magnet. Which sample as a whole would be better for making a large magnet?

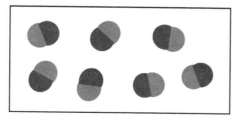

Sample A

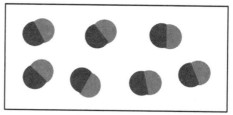

Sample B

Explain your answer. Why did you make that decision?

3. In science, laws are more important than theories because laws are true.

    a. I agree with this statement.

    b. I disagree with this statement.

    Explain your answer, using an example from your investigation about magnetic force.

National Science Teachers Association

4. In science, evidence for our claims comes from the data we collect.

   a. I agree with this statement.
   b. I disagree with this statement.

   Explain your answer, using an example from your investigation about magnetic force.

5. The amount of matter and energy in the universe is constant. Explain how understanding the movement of energy and matter within and between systems is helpful to scientists. Use an example from your investigation about magnetic force to help in your explanation.

6. The structure of an object is usually related to the function of that object. Use an example from your investigation about magnetic force to explain why it is important to understand the relationship between structure and function within science.

# Application Labs

## Lab Handout

# Lab 11. Design Challenge
## Which Electromagnet Design Is Best for Picking Up 50 Paper Clips?

**Electromagnet made from a D-cell battery, a copper wire, and an iron nail**

### Introduction

Electromagnets are formed when a wire is wrapped around a metal core and electricity flows through the wire. When electricity flows through the wire, a magnetic field is produced, which is then concentrated in the metal core. It is possible to make a simple electromagnet by wrapping copper wire around an iron rod or nail and then connecting the wire to a small battery (see Figure L11.1). When electricity from the battery flows through the wire, the magnetic field from the electric current causes the iron core to have magnetic properties.

The iron core of an electromagnet is a key component for making the device work. The atoms of iron contain a large number of electrons. Within the iron core, the atoms and their electrons are moving about in random directions. However, when the atoms are exposed to the magnetic field from the electric current in the wire, the atoms and electrons change their orientations and movement to align with the magnetic field of the electric current. This rearrangement of the atoms amplifies the magnetic field from the wire and results in a stronger electromagnet. When the electric current stops (by turning it off or disconnecting the battery), the atoms in the iron core go back to their random motion and positions and no longer have magnetic properties.

There are three main parts required to build an electromagnet: the iron core, copper wire, and an electricity source. Changes in each of these pieces of the electromagnet will influence the overall strength of the magnet. The more powerful the source of electricity, the more electric current there will be flowing through the wire. The larger the iron core, the more atoms there are that can be aligned to help amplify the magnetic force. The number of times the wire is wrapped around the iron rod will also influence the strength of the electromagnet.

Electromagnets are used in many different contexts, ranging from small applications, such as the audio speakers in a cell phone, to large applications, such as picking up old cars in a junkyard. In each of these contexts, the electromagnet must be strong enough to accomplish its intended purpose but not so strong that it has unintended effects on nearby materials. In some cases, an electromagnet may be used to pick up a specific weight or even a specific amount of objects. If an electromagnet is designed for a specific task, it must have the right combination of properties, including the amount of electricity, the size of the iron core, and the number of times the wire is wrapped around the iron core so that it has a consistent and predictable performance.

## Your Task

Use what you know about magnetic fields and cause-and-effect relationships to design a small electromagnet that can be used for a specific task. Your specific design challenge is as follows: The International Paperclip Company (IPC) needs an electromagnet that will pick up approximately 50 paper clips to help fill and package a single box of clips. For the IPC to keep their customers happy and the company's costs consistent, the electromagnet must reliably pick up between 47 and 53 paper clips. The IPC would also like to spend as little money as possible to construct their new electromagnet, while making sure that their new electromagnet will accomplish its task.

The guiding question of this investigation is, **Which electromagnet design is best for picking up 50 paper clips?**

## Materials

You may use any of the following materials during your investigation:

- Batteries (different sizes)
- Copper wire
- Iron nails (different sizes)
- Paperclips
- Gauss meter (optional)
- Safety glasses or goggles

## Safety Precautions

Follow all normal lab safety rules. In addition, take the following safety precautions:

1. Wear sanitized safety glasses or goggles during lab setup, hands-on activity, and takedown.

2. Given that the electromagnet setup will generate heat, disconnect the wire from the battery when not actively collecting data, to prevent skin burn.

3. Use caution when wrapping the copper wire around the nail. It is possible that the loose end will swing around during the wrapping process and present a hazard.

4. Use caution in handling the wire end or nails. They are sharp and can cut or puncture skin.

5. Wash hands with soap and water after completing the lab activity.

## Investigation Proposal Required?     ☐ Yes          ☐ No

## Getting Started

To answer the guiding question, you will need to design and test an electromagnet that is capable of picking up 50 paper clips. You must also consider the cost of the materials that are used in your electromagnet. You will need to systematically test how changes to each part of the electromagnet influence the strength of the magnet and its ability to accomplish its intended task. To complete this design challenge, you must determine what type of data you need to collect, how you will collect it, and how you will analyze it.

To determine *what type of data you need to collect,* think about the following questions:

- How will you determine the strength of the electromagnet?
- What information or measurements will you need to record?
- What parts of the electromagnet will you change and what parts will you keep consistent?
- What design factors will you consider when building your electromagnet?

To determine *how you will collect your data,* think about the following questions:

- What equipment will you need to collect the data you need?
- How will you make sure that your data are of high quality (i.e., how will you reduce error)?
- How will you keep track of the data you collect?
- How will you organize your data?

To determine *how you will analyze your data,* think about the following questions:

- How will you determine which parts of the electromagnet have the biggest impact on its strength?
- What type of table or graph could you create to help make sense of your data?
- How will you determine if one electromagnet design is better than another?

When you design your electromagnet for the IPC, it will be important to take into account the amount of materials you use, how much they cost, and how often the components will need to be replaced. Table L11.1 includes information about the different materials that might be used in making your electromagnet.

## TABLE L11.1

**Cost information for electromagnet materials**

| Item | Cost |
|------|------|
| Battery, 1.5 v, size D | $0.99/each |
| Battery, 1.5 v, size C | $0.89/each |
| Battery, 1.5 v, size AA | $0.45/each |
| Copper wire | $0.07/30 cm |
| Iron core | $0.015/cm |

## Connections to Crosscutting Concepts, the Nature of Science, and the Nature of Scientific Inquiry

As you work through your investigation, be sure to think about

- why it is important to understand cause-and-effect relationships,
- how defining a system under study and making a model of it is helpful in science,
- how scientists use different methods to answer different types of questions, and
- the role of imagination and creativity in science.

## Initial Argument

Once your group has finished collecting and analyzing your data, your group will need to develop an initial argument. Your initial argument needs to include a *claim*, *evidence* to support your claim, and a *justification* of the evidence. The claim is your group's answer to the guiding question. The evidence is an analysis and interpretation of your data. Finally, the justification of the evidence is why your group thinks the evidence matters. The justification of the evidence is important because scientists can use different kinds of evidence to support their claims. Your group will create your initial argument on a whiteboard. Your whiteboard should include all the information shown in Figure L11.2.

**FIGURE L11.2** _____

**Argument presentation on a whiteboard**

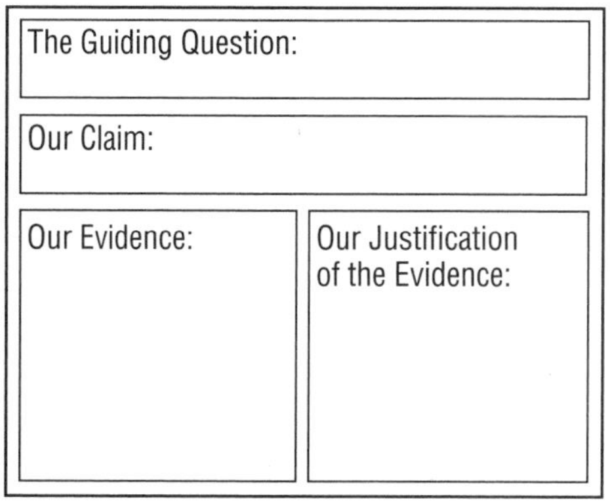

| The Guiding Question: | |
|---|---|
| Our Claim: | |
| Our Evidence: | Our Justification of the Evidence: |

## Argumentation Session

The argumentation session allows all of the groups to share their arguments. One member of each group will stay at the lab station to share that group's argument, while the other members of the group go to the other lab stations to listen to and critique the arguments developed by their classmates. This is similar to how scientists present their arguments to other scientists at conferences. If you are responsible for critiquing your classmates' arguments, your goal is to look for mistakes so these mistakes can be fixed and they can make their argument better. The argumentation session is also a good time to think about ways you can make your initial argument better. Scientists must share and critique arguments like this to develop new ideas.

To critique an argument, you might need more information than what is included on the whiteboard. You will therefore need to ask the presenter lots of questions. Here are some good questions to ask:

- How did you collect your data? Why did you use that method? Why did you collect those data?

- What did you do to make sure the data you collected are reliable? What did you do to decrease measurement error?

- How did you group analyze the data? Why did you decide to do it that way? Did you check your calculations?

- Is that the only way to interpret the results of your analysis? How do you know that your interpretation of your analysis is appropriate?

- Why did your group decide to present your evidence in that way?

- What other claims did your group discuss before you decided on that one? Why did your group abandon those alternative ideas?

- How confident are you that your claim is valid? What could you do to increase your confidence?

Once the argumentation session is complete, you will have a chance to meet with your group and revise your initial argument. Your group might need to gather more data or design a way to test one or more alternative claims as part of this process. Remember, your goal at this stage of the investigation is to develop the most acceptable and valid answer to the research question!

## Report

Once you have completed your research, you will need to prepare an *investigation report* that consists of three sections. Each section should provide an answer to the following questions:

1. What question were you trying to answer and why?

2. What did you do to answer your question and why?

3. What is your argument?

Your report should answer these questions in two pages or less. This report must be typed, and any diagrams, figures, or tables should be embedded into the document. Be sure to write in a persuasive style; you are trying to convince others that your claim is acceptable and valid!

*Checkout Questions*

# Lab 11. Design Challenge
## Which Electromagnet Design Is Best for Picking Up 50 Paper Clips?

1. A compass needle is a small magnet that aligns with the magnetic field of Earth and is used to help people tell which direction is north when they are traveling. But whenever a compass gets close to a strong electric current, the compass needle points in a different direction—see the drawings below.

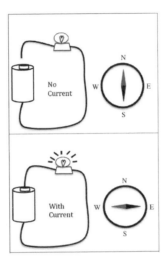

   Use the images and what you know about magnets and electric currents to explain why the compass needle will move when there is an electric current present.

# LAB 11

2. Look at the drawings of the electromagnets below. Which one will be the strongest—A, B, C, or D?

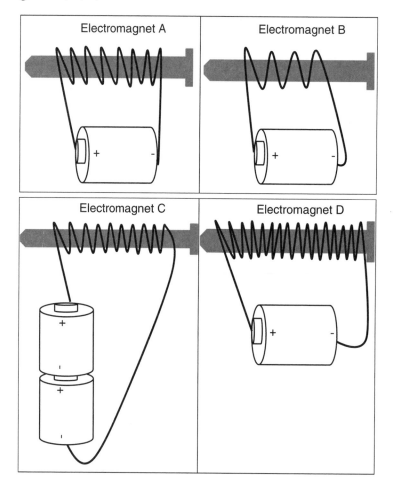

Explain your answer. Why did you make that decision?

3. When scientists use their imagination and creativity, it makes their work less scientific.

    a. I agree with this statement.

    b. I disagree with this statement.

Explain your answer, using an example from your experience designing an electromagnet.

4. Scientists use the scientific method to design investigations to answer questions.

    a. I agree with this statement.

    b. I disagree with this statement.

Explain your answer, using an example from your experience designing an electromagnet.

5. Changing one aspect of a system or one variable in an investigation can have an impact on many other things. Explain how understanding cause-and-effect relationships is helpful to scientists. Use an example from your experience designing an electromagnet to help in your explanation.

6. Before designing equipment, engineers usually make mathematical models or smaller versions of the equipment to test out how it will work. Using an example from your experience designing an electromagnet, explain why it is important for scientists and engineers to make models of systems.

*Lab Handout*

# Lab 12. Unbalanced Forces
## How Does Surface Area Influence Friction and the Motion of an Object?

### Introduction

The motion of an object is determined by the combination of all the forces acting on that object. Those forces may come in the form of a simple push or pull that result from contact forces, such as pushing a box with your hand, or non-contact forces, such as gravity pulling an object toward Earth. When forces are acting on an object from the same direction, the influence of each force is added together. Think of two people pushing a box across the floor. If they push from the same side, the forces acting on the box are unbalanced, meaning that there is more force on one side of the box than the other. When the forces acting on the box are unbalanced, the box moves in the direction of that force. However, if the two people stand on opposite sides of the box and push again, their forces are working against each other (see Figure L12.1). One person is trying to push the box to the right and the other person is trying to push the box to the left. In this case, their forces are balanced and cancel each other out. When the forces acting on an object are balanced, the motion of the object does not change.

## FIGURE L12.1 _____

**Two people pushing on opposite sides of a box**

When two people push on a box in opposite directions, it's easy to see that there are opposing forces acting on the box, but sometimes there are opposing forces acting on an object that are less obvious. If you give a book a push and let it slide across a table, it will eventually come to a stop. The book stops because there is an opposite force acting on it, just like the person pushing on the opposite side of the box. The force that causes the book to come to a stop is called friction. Friction is a force that occurs when two surfaces are in contact with each other. The force of friction always works on an object in a direction that is opposite of the motion of the object. Therefore, friction is always trying to slow an object down or keep an object from moving. Trying to slide a heavy object is often difficult due to friction. Placing a heavy object on a cart with wheels can make it easier to move. The amount of friction is greatly reduced by the wheels, so it is easier to move the heavy object.

The amount of friction between two objects depends on several factors. One important factor is the specific surfaces involved. The amount of friction between two objects as they slide past one another depends on the specific surfaces that are rubbing against each other. For example, a cardboard box sliding on a wooden floor has a different amount of friction than the same cardboard box sliding on a carpet floor. Different combinations of surface materials will result in different amounts of friction. Another factor that influences the

# LAB 12

amount of friction is the weight of the object that is moving. The heavier the object, the more friction there will be that opposes the motion of the object.

Reducing friction is often an important goal when there is work to be done. There are many different ways that someone can try to reduce the amount of friction when moving an object. One strategy is to change the surface that an object is sliding on. For example, there is very little friction between ice and other materials, therefore objects slide over ice very easily. Another strategy for reducing the amount of friction when trying to slide an object would be reducing the mass of the object; then there would be less force between the moving object and the surface it is sliding on. A third strategy would be to change the shape of the object and change the amount of contact between the two surfaces that are rubbing together. For example, instead of sliding two separate boxes side by side, perhaps stacking them one on top of the other would reduce the amount of friction and make them easier to move.

## Your Task

Use what you know about forces and motion, patterns, and stability and change to design and carry out an investigation that will allow you to test how changing the amount of surface area between two objects influences the amount of friction between the objects and how changing the amount of surface area influences the motion of the object.

The guiding question of this investigation is, **How does surface area influence friction and the motion of an object?**

## Materials

You may use any of the following materials during your investigation:

- Friction block set
- 20 N spring scale
- 10 N spring scale
- 5 N spring scale
- Mass set
- Meterstick
- Stopwatch
- Safety glasses or goggles

## Safety Precautions

Follow all normal lab safety rules. In addition, take the following safety precautions:

1. Wear sanitized safety glasses or goggles during lab setup, hands-on activity, and takedown.

2. Wash hands with soap and water after completing the lab activity.

National Science Teachers Association

**Investigation Proposal Required?**  ☐ Yes  ☐ No

## Getting Started

To answer the guiding question, you will need to design and conduct an investigation to measure the amount of friction between the friction block and the surface of your table and test how changes in the amount of surface area influence the motion of the block. To accomplish this task, you must determine what type of data you need to collect, how you will collect it, and how you will analyze it.

To determine *what type of data you need to collect*, think about the following questions:

- How will you determine the force due to friction?
- What information or measurements will you need to record?
- How will you determine the surface area of the block?
- Will you test different surface combinations?

To determine *how you will collect your data*, think about the following questions:

- What equipment will you need to collect the data you need?
- How will you make sure that your data are of high quality (i.e., how will you reduce error)?
- How will you keep track of the data you collect?
- How will you organize your data?

To determine *how you will analyze your data*, think about the following questions:

- How will you determine the influence of surface area on friction and motion of the block?
- What type of table or graph could you create to help make sense of your data?

## Connections to Crosscutting Concepts, the Nature of Science, and the Nature of Scientific Inquiry

As you work through your investigation, be sure to think about

- why it is important to understand patterns and their causes,
- the importance of understanding stability and change within systems,
- the difference between observations and inferences, and
- the role of different methods in science.

## Initial Argument

Once your group has finished collecting and analyzing your data, your group will need to develop an initial argument. Your initial argument needs to include a *claim, evidence* to support your claim, and a *justification* of the evidence. The claim is your group's answer to the guiding question. The evidence is an analysis and interpretation of your data. Finally, the justification of the evidence is why your group thinks the evidence matters. The justification of the evidence is important because scientists can use different kinds of evidence to support their claims. Your group will create your initial argument on a whiteboard. Your whiteboard should include all the information shown in Figure L12.2.

## FIGURE L12.2

**Argument presentation on a whiteboard**

| The Guiding Question: | |
|---|---|
| Our Claim: | |
| Our Evidence: | Our Justification of the Evidence: |

## Argumentation Session

The argumentation session allows all of the groups to share their arguments. One member of each group will stay at the lab station to share that group's argument, while the other members of the group go to the other lab stations to listen to and critique the arguments developed by their classmates. This is similar to how scientists present their arguments to other scientists at conferences. If you are responsible for critiquing your classmates' arguments, your goal is to look for mistakes so these mistakes can be fixed and they can make their argument better. The argumentation session is also a good time to think about ways you can make your initial argument better. Scientists must share and critique arguments like this to develop new ideas.

To critique an argument, you might need more information than what is included on the whiteboard. You will therefore need to ask the presenter lots of questions. Here are some good questions to ask:

- How did you collect your data? Why did you use that method? Why did you collect those data?

- What did you do to make sure the data you collected are reliable? What did you do to decrease measurement error?

- How did your group analyze the data? Why did you decide to do it that way? Did you check your calculations?

- Is that the only way to interpret the results of your analysis? How do you know that your interpretation of your analysis is appropriate?

- Why did your group decide to present your evidence in that way?

- What other claims did your group discuss before you decided on that one? Why did your group abandon those alternative ideas?

- How confident are you that your claim is valid? What could you do to increase your confidence?

Once the argumentation session is complete, you will have a chance to meet with your group and revise your initial argument. Your group might need to gather more data or design a way to test one or more alternative claims as part of this process. Remember, your goal at this stage of the investigation is to develop the most acceptable and valid answer to the research question!

## Report

Once you have completed your research, you will need to prepare an *investigation report* that consists of three sections. Each section should provide an answer to the following questions:

1. What question were you trying to answer and why?

2. What did you do to answer your question and why?

3. What is your argument?

Your report should answer these questions in two pages or less. This report must be typed, and any diagrams, figures, or tables should be embedded into the document. Be sure to write in a persuasive style; you are trying to convince others that your claim is acceptable and valid!

# LAB 12

## Checkout Questions

# Lab 12. Unbalanced Forces
## How Does Surface Area Influence Friction and the Motion of an Object?

1. Jared knows that trucks and cars driving down the road normally have good traction between the rubber tires and the asphalt of the road. But he is unsure why trucks and cars begin to slide if they hit a patch of ice. Using what you know about balanced and unbalanced forces, explain to Jared why truck and car wheels slide on ice rather than roll like they do on normal roads.

2. A student in physical science class was conducting an investigation by sliding different blocks across her wooden lab table. Each block was launched by a rubber band with the same force. She measured how far the block traveled and obtained the following results:

| Surface of block | Distance traveled | | |
|---|---|---|---|
| | Trial 1 | Trial 2 | Trial 3 |
| Wood | 57 cm | 66 cm | 52 cm |
| Aluminum foil | 64 cm | 71 cm | 69 cm |
| Carpet | 32 cm | 40 cm | 45 cm |

Use what you know about balanced and unbalanced forces to generate an argument (including a claim, evidence, and justification) that explains the results the student obtained.

National Science Teachers Association

3. When scientists make observations, they are more certain than when they make inferences.

    a.  I agree with this statement.
    b.  I disagree with this statement.

    Explain your answer, using an example from your investigation on unbalanced forces.

4. When several scientists are investigating the same thing, they all use the same methods so that they get the same answer.

    a.  I agree with this statement.
    b.  I disagree with this statement.

    Explain your answer, using an example from your investigation on unbalanced forces.

5. Scientists often look for patterns in nature or within the data that they collect during an investigation. Using an example from your investigation about unbalanced forces, explain why it is important to understand patterns and their causes within science.

6. In physics, there are times when scientists study many variables to learn how a system works. Using an example from your investigation on unbalanced forces, explain why it is important for scientists to understand what causes systems to be stable or change.

# SECTION 4
# Physical Science
# Core Idea 3

## Energy

# Introduction Labs

# Lab 13. Kinetic Energy
## How Do the Mass and Velocity of an Object Affect Its Kinetic Energy?

### Introduction

When law enforcement officials investigate car crashes (see Figure L13.1), it can sometimes be difficult to determine who is at fault and what laws were broken, especially when there is no footage of the crash. To re-create the crash scene, investigators use physics concepts to determine the specifics of a crash, including the speed and direction (together, the velocity) in which a car was traveling and when the driver attempted to stop. These figures can then be used to help determine who is at fault in a crash and the laws that person broke.

## FIGURE L13.1

**A police officer investigates a crash scene.**

You already know that energy is conserved and transferred within and between systems, not created or destroyed. This is also true in car crashes. A traveling car has a certain amount of kinetic energy, and when that car hits another car or a different object, some of that energy is transformed into heat or sound, but most is used to do the work that deforms the car or object it crashes into. So, when cars collide, the transfer of their kinetic energy is responsible for the resulting damage. The damage done to the cars can be used, along with other pieces of evidence, to determine the velocity of the car.

In this investigation, you will be applying the same physics concepts to determine the mass or velocity of an object that has been dropped onto an inelastic surface (one that does not return to the same shape after impact). You will be creating a collision between a ball of variable mass and a large container of flour to investigate this relationship.

### Your Task

Use what you know about force and motion, patterns, and causal relationships to design and carry out an investigation that will allow you to create a mathematical model explaining the relationship between mass, velocity, and force of impact.

The guiding question of this investigation is, **How do the mass and velocity of an object affect its kinetic energy?**

## Materials

You may use any of the following materials during your investigation:

- Large plastic container filled with flour
- Racquetball with slit
- Funnel
- Rice or beans
- Meterstick
- Ruler
- String
- Stopwatch
- Electronic or triple beam balance
- Excel, graphing calculator, or other mathematical software (optional)
- Safety glasses or goggles

## Safety Precautions

Follow all normal lab safety rules. In addition, take the following safety precautions:

1. Wear sanitized indirectly vented chemical-splash safety goggles during lab setup, hands-on activity, and takedown.

2. Never put consumables in your mouth.

3. Sweep up flour off the floor to avoid a slip or fall hazard.

4. Do not throw objects.

5. Wash hands with soap and water after completing the lab activity.

## Investigation Proposal Required?    ☐ Yes    ☐ No

## Getting Started

The first step in developing your mathematical model is to design and carry out an investigation to determine how mass and velocity affect the kinetic energy of the variable-mass ball. Because the kinetic energy of the ball will deform the flour surface during an impact, the extent of this deformation can be used to determine the kinetic energy of the ball at impact.

To determine *what type of data you need to collect*, think about the following questions:

- What information do you need to create your mathematical model?
- What measurements will you take during your investigation?
- How will you determine the velocity of the ball?
- In what way will you vary the mass of the ball, if at all?
- How will you know how much kinetic energy has been absorbed by the flour surface?

To determine *how you will collect your data*, think about the following questions:

- What equipment will you need to collect the data you need?

- How will you make sure that your data are of high quality (i.e., how will you reduce error)?
- How will you keep track of the data you collect?
- How will you organize your data?

To determine *how you will analyze your data,* think about the following questions:

- What type of calculations will you need to make?
- What type of table or graph could you create to help make sense of your data?

Once you have carried out your investigations, your group will need to use the data you collected to develop a mathematical model that can be used to help explain the relationship between the mass of an object, the velocity of the object, and the kinetic energy of the object. You should be able to use your mathematical model to predict any one of the three values when given the other two. For example, if given the kinetic energy measurement and the object's mass, you should be able to use your mathematical model to determine the velocity of the object at impact.

The last step in this investigation is to test your mathematical model. To accomplish this goal, you can predict the kinetic energy measurement of a ball with a given mass and velocity. If you are able to use your mathematical model to make accurate predictions about the object's mass, velocity, and kinetic energy, you will be able to generate the evidence you need to convince others that the mathematical model you developed is valid.

## Connections to Crosscutting Concepts, the Nature of Science, and the Nature of Scientific Inquiry

As you work through your investigation, be sure to think about

- the importance of identifying cause and effect;
- how scientists often need to track how energy moves into, out of, and within a system;
- the difference between laws and theories in science; and
- how scientists must use imagination and creativity when developing models and explanations.

## Initial Argument

Once your group has finished collecting and analyzing your data, your group will need to develop an initial argument. Your initial argument needs to include a *claim, evidence* to support your claim, and a *justification* of the evidence. The claim is your group's answer to the guiding question. The evidence is an analysis and interpretation of your data. Finally, the justification of the evidence is why your group thinks the evidence matters. The justification of the evidence is important because scientists can use different kinds of evidence to support their claims. Your group will create your initial argument on a whiteboard. Your whiteboard should include all the information shown in Figure L13.2.

## Argumentation Session

The argumentation session allows all of the groups to share their arguments. One member of each group will stay at the lab station to share that group's argument, while the other members of the group go to the other lab stations to listen to and critique the arguments developed by their classmates. This is similar to how scientists present their arguments to other scientists at conferences. If you are responsible for critiquing your classmates' arguments, your goal is to look for mistakes

**FIGURE L13.2**

**Argument presentation on a whiteboard**

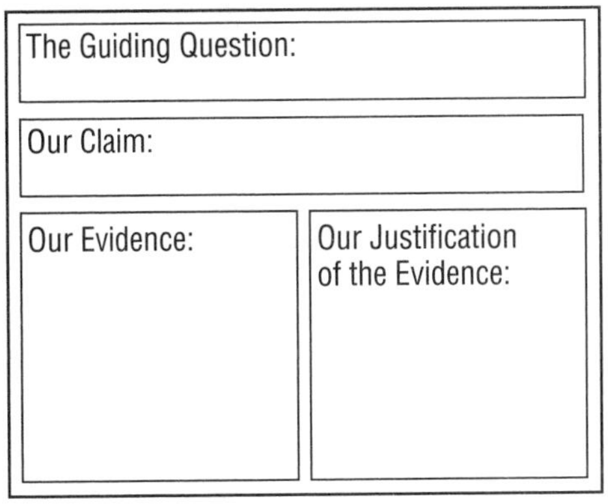

| The Guiding Question: | |
|---|---|
| Our Claim: | |
| Our Evidence: | Our Justification of the Evidence: |

so these mistakes can be fixed and they can make their argument better. The argumentation session is also a good time to think about ways you can make your initial argument better. Scientists must share and critique arguments like this to develop new ideas.

To critique an argument, you might need more information than what is included on the whiteboard. You will therefore need to ask the presenter lots of questions. Here are some good questions to ask:

- How did you collect your data? Why did you use that method? Why did you collect those data?

- What did you do to make sure the data you collected are reliable? What did you do to decrease measurement error?

- How did your group analyze the data? Why did you decide to do it that way? Did you check your calculations?

- Is that the only way to interpret the results of your analysis? How do you know that your interpretation of your analysis is appropriate?

- Why did your group decide to present your evidence in that way?

- What other claims did your group discuss before you decided on that one? Why did your group abandon those alternative ideas?

- How confident are you that your claim is valid? What could you do to increase your confidence?

Once the argumentation session is complete, you will have a chance to meet with your group and revise your initial argument. Your group might need to gather more data or design a way to test one or more alternative claims as part of this process. Remember, your goal at this stage of the investigation is to develop the most acceptable and valid answer to the research question!

# LAB 13

## Report

Once you have completed your research, you will need to prepare an *investigation report* that consists of three sections. Each section should provide an answer to the following questions:

1. What question were you trying to answer and why?

2. What did you do to answer your question and why?

3. What is your argument?

Your report should answer these questions in two pages or less. This report must be typed, and any diagrams, figures, or tables should be embedded into the document. Be sure to write in a persuasive style; you are trying to convince others that your claim is acceptable and valid!

*Checkout Questions*

# Lab 13. Kinetic Energy
## How Do the Mass and Velocity of an Object Affect Its Kinetic Energy?

1. Malik is throwing rocks into a lake. He throws a 1 kg rock that travels 5 m/s immediately after it left his hand. He also throws a 0.5 kg rock that travels 8 m/s immediately after it left his hand. Which rock did Malik throw the hardest?

   Explain your answer. How do you know which rock Malik threw the hardest?

2. Jerilyn dropped three spheres with masses of 0.25 kg, 0.75 kg, and 1 kg from equal height into a tub of flour. After dropping the spheres, her lab partner, Evan, put the spheres away before she recorded her data. Jerilyn and Evan are now unsure which sphere created which crater. A side view of the craters is shown below. Use what you know about mass, velocity, and kinetic energy to select which sphere created each crater.

Explain your answer. How do you know which sphere made which crater? Use examples from your investigation about kinetic energy to support your answer.

3. Thinking creatively in science will lead to work that is less scientific and valid.

   a. I agree with this statement.

   b. I disagree with this statement.

   Explain your answer, using an example from your investigation about kinetic energy.

4. In science, theories and laws describe the same thing.

   a. I agree with this statement.

   b. I disagree with this statement.

   Explain your answer, using an example from your investigation about kinetic energy.

5. Understanding the ways in which the transfer of energy affects objects within a system is an important aspect of science and engineering. Using an example from your investigation about kinetic energy, explain why tracking the transfer of energy within a system is important.

6. Identifying cause-and-effect relationships in nature can help scientists make predictions about the behavior of objects. What cause-and-effect relationships did you observe in your investigation about kinetic energy, and how do these relationships allow you to make accurate predictions?

## Lab Handout

# Lab 14. Potential Energy
## How Can You Make an Action Figure Jump Higher?

### Introduction

Teeterboards are typical pieces of equipment found on many playgrounds around the country. They are often used in shows that focus on gymnastic tricks. The picture in Figure L14.1 shows a circus act involving a performer launching another performer high into the air. It is easy to observe how the activity of a teeterboard involves objects' motion. However, that activity also involves energy shifting between forms.

The law of conservation of energy states that within a given system the total amount of energy always stays the same—it is neither created nor destroyed; instead, energy is transformed from one form to another. When energy is stored in one form or another, it is called potential energy. Potential energy can be stored in the chemical bonds between atoms in a molecule and in the nuclei of atoms. Energy can also be stored based on the position of an object. Indeed, potential energy can be referred to as energy of position. When potential energy is transformed into motion, it becomes kinetic energy. Kinetic energy can be detected when objects move. Kinetic energy is known as energy of motion.

For an example, think about climbing a hill. When you are at the bottom of a hill, you have low potential energy based on your position. To increase your potential energy, you climb to the top of the hill. As you are climbing, you are moving, so you are using kinetic energy; you are transforming kinetic energy into increased potential energy; and you are changing position. Since you have climbed higher, you have greater potential energy. In this investigation you will explore the relationship between potential energy and kinetic energy as you try to make an action figure jump using a teeterboard.

## FIGURE L14.1
**Circus performers on a teeterboard**

### The Task

Use what you know about the conservation of energy and models to design and carry out an investigation that will allow you to develop a rule that explains how an action figure can be made to jump lower or higher on a teeterboard.

The guiding question of this investigation is, **How can you make an action figure jump higher?**

## Materials

You may use any of the following materials during your investigation:

- Ruler
- Meterstick
- Electronic or triple beam balance
- Pencil

- Clay (100 g)
- Action figures
- Safety glasses or goggles

## Safety Precautions

Follow all normal lab safety rules. In addition, take the following safety precautions:

1. Wear sanitized safety glasses or goggles during lab setup, hands-on activity, and takedown.

2. Sweep clay up off the floor to avoid a slip or fall hazard.

3. Do not allow the action figure to jump too far from your work area.

4. Remove any fragile items from the work area.

5. Wash hands with soap and water after completing the lab activity.

## Investigation Proposal Required?     ☐ Yes      ☐ No

## Getting Started

To answer the guiding question, you will need to design and conduct an investigation that explores changing the potential energy of an action figure. To accomplish this task, you must determine what type of data you need to collect, how you will collect it, and how you will analyze it.

To determine *what type of data you need to collect*, think about the following questions:

- How will you test the ability to make the action figure jump higher?
- How will you measure the height of the jump?
- What type of measurements or observations will you need to record during your investigation?

To determine *how you will collect your data*, think about the following questions:

- How often will you collect data and when will you do it?

- How will you make sure that your data are of high quality (i.e., how will you reduce error)?

- How will you keep track of the data you collect and how will you organize it?

To determine *how you will analyze your data,* think about the following questions:

- What type of calculations will you need to make?

- What type of graph could you create to help make sense of your data?

## Connections to Crosscutting Concepts, the Nature of Science, and the Nature of Scientific Inquiry

As you work through your investigation, be sure to think about

- how defining systems and models provides tools for understanding and testing of ideas;

- why it is important to track how energy and matter flows into, out of, and within a system;

- the difference between laws and theories in science; and

- the different forms of scientific investigation, including experiments, systematic observations, and analysis of data sets.

## Initial Argument

Once your group has finished collecting and analyzing your data, your group will need to develop an initial argument. Your initial argument needs to include a *claim, evidence* to support your claim, and a *justification* of the evidence. The claim is your group's answer to the guiding question. The evidence is an analysis and interpretation of your data. Finally, the justification of the evidence is why your group thinks the evidence matters. The justification of the evidence is important because scientists can use different kinds of evidence to support their claims. Your group will create your initial argument on a whiteboard. Your whiteboard should include all the information shown in Figure L14.2.

## FIGURE L14.2 _____

**Argument presentation on a whiteboard**

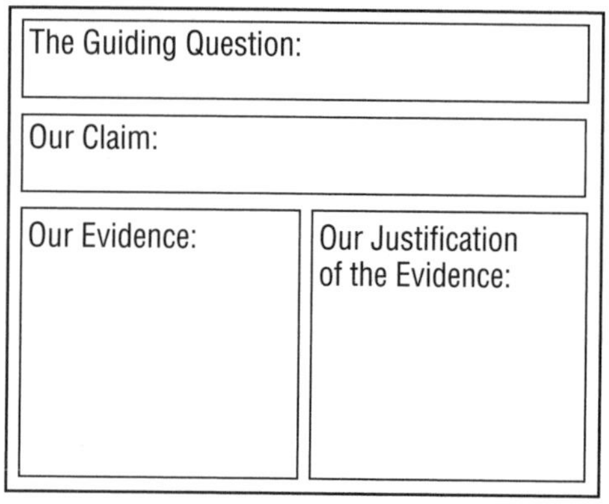

| The Guiding Question: |  |
|---|---|
| Our Claim: | |
| Our Evidence: | Our Justification of the Evidence: |

## Argumentation Session

The argumentation session allows all of the groups to share their arguments. One member of each group will stay at the lab station to share that group's argument, while the other members of the group go to the other lab stations to listen to and critique the arguments developed by their classmates. This is similar to how scientists present their arguments to other scientists at conferences. If you are responsible for critiquing your classmates' arguments, your goal is to look for mistakes so these mistakes can be fixed and they can make their argument better. The argumentation session is also a good time to think about ways you can make your initial argument better. Scientists must share and critique arguments like this to develop new ideas.

To critique an argument, you might need more information than what is included on the whiteboard. You will therefore need to ask the presenter lots of questions. Here are some good questions to ask:

- How did you collect your data? Why did you use that method? Why did you collect those data?
- What did you do to make sure the data you collected are reliable? What did you do to decrease measurement error?
- How did your group analyze the data? Why did you decide to do it that way? Did you check your calculations?
- Is that the only way to interpret the results of your analysis? How do you know that your interpretation of your analysis is appropriate?
- Why did your group decide to present your evidence in that way?
- What other claims did your group discuss before you decided on that one? Why did your group abandon those alternative ideas?
- How confident are you that your claim is valid? What could you do to increase your confidence?

Once the argumentation session is complete, you will have a chance to meet with your group and revise your initial argument. Your group might need to gather more data or design a way to test one or more alternative claims as part of this process. Remember, your goal at this stage of the investigation is to develop the most acceptable and valid answer to the research question!

## Report

Once you have completed your research, you will need to prepare an *investigation report* that consists of three sections. Each section should provide an answer to the following questions:

1. What question were you trying to answer and why?
2. What did you do to answer your question and why?
3. What is your argument?

Your report should answer these questions in two pages or less. This report must be typed, and any diagrams, figures, or tables should be embedded into the document. Be sure to write in a persuasive style; you are trying to convince others that your claim is acceptable and valid!

*Checkout Questions*

# Lab 14. Potential Energy
## How Can You Make an Action Figure Jump Higher?

1. What is potential energy?

2. What is kinetic energy?

3. A student is trying to get a cart to reach the wall at the end of the system pictured below. He uses a ramp to get the cart some energy to cover that distance. However, as shown below, using the ramp as constructed, he was not able to reach the wall.

   a. What can the student change to get the cart to reach the wall?

   b. How do you know?

4. The law of conservation of energy describes how energy exists in physical systems but not why it acts in certain ways.

   a. I agree with this statement.

   b. I disagree with this statement.

   Explain your answer, using an example from your investigation about potential energy.

5. Science only relies on experiments to understand the physical world.

   a. I agree with this statement.

   b. I disagree with this statement.

   Explain your answer, using an example from your investigation about potential energy.

6. Scientists often have to define the boundaries of physical systems and use them to create models to test ideas. Explain why defining systems and models is important in science, using an example from your investigation about potential energy.

7. It is important to track how energy flows into, out of, and within a system during an investigation. Explain why it is important to keep track of energy when studying a system, using an example from your investigation about potential energy.

# LAB 15

## Lab 15. Thermal Energy and Specific Heat
### Which Material Has the Greatest Specific Heat?

### Introduction

Scientists are able to identify unknown substances based on their chemical and physical properties. A substance is a type of matter with a specific composition and specific properties. One physical property of a substance is the amount of energy it will absorb per unit of mass. This property is called specific heat ($s$). Specific heat is the amount of energy, measured in joules, that is needed to raise the temperature of 1 gram of the substance 1 degree Celsius. Scientists often need to know the specific heat of different substances when they attempt to track how energy moves into, out of, and within a system.

Chemists use a technique called calorimetry to determine the specific heat of a substance. Calorimetry, or the measurement of heat transfer, is based on the law of conservation of energy. This law states that energy is not created nor destroyed; it is only converted from one form to another. This fundamental law serves as the foundation for all the research that is done in the field of thermodynamics, which is the study of heat, temperature, and heat transfer. Heat is defined as the total kinetic energy of all the atoms or molecules that make up a substance. Temperature, in contrast, is defined as a measure of the average kinetic energy of the atoms or molecules that make up a substance.

Heat, or thermal energy, can be transferred through a substance and between two different objects. Scientists call this process conduction (see Figure L15.1). The transfer of heat energy through the process of conduction can be explained by thinking of the heat from a source causing the atoms of a substance to vibrate faster, which means they have greater kinetic energy. These atoms then cause the atoms next to them to vibrate faster by bumping into them, which means that the kinetic energy of the neighboring atoms increases as well. Over time, kinetic energy is transferred from one atom to the next. As more atoms in the substance gain kinetic energy over time, the temperature of the substance increases. This process is also how heat energy is able to transfer between two different objects that are in contact with each other.

The amount of heat ($q$) transferred to an object depends on three factors. The first is the mass ($m$) of the object. The second factor is the specific heat ($s$) value of object. This is important because an object will consist of a specific type of substance, and each type of substance has a unique specific heat value. The third factor is the resulting temperature change ($\Delta T$). The mathematical relationship between these three factors and the amount of heat transferred to an object is

$$q = m \times s \times \Delta T$$

# FIGURE L15.1

**Thermal energy can transfer through a substance or from one substance to another by conduction.**

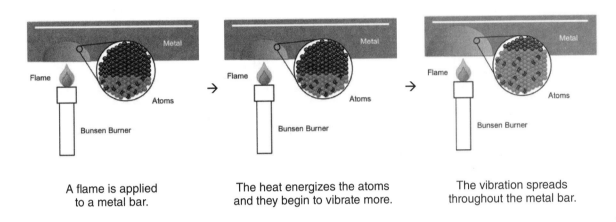

A flame is applied to a metal bar.

The heat energizes the atoms and they begin to vibrate more.

The vibration spreads throughout the metal bar.

The materials that people use to build a new structure or to manufacture commercial goods have a wide range of specific heat values. Take concrete and wood as an example. Both of these materials can be used to build benches in parks or at bus stops for people to use. Wood, however, has a much higher specific heat than concrete. It therefore takes more heat energy to increase the temperature of a 10 kg piece of wood than it does to increase the temperature of a 10 kg piece of concrete. The piece of concrete, as a result, will get hotter faster than the piece of wood when it is exposed to the same amount of heat energy. This issue could be a potential problem in cities that tend to be hot and sunny most of the year. Engineers and manufacturers therefore need to know how to look up or determine the specific heat value of a potential building or manufacturing material before they decide to use it. In this investigation, you will have an opportunity to learn how to determine the specific heat value of a material using the process of calorimetry.

## Your Task

Use what you know about heat, temperature, the conservation of energy, and defining systems to design and carry out an investigation to determine the specific heat values of several different materials.

The guiding question of this investigation is, **Which material has the greatest specific heat?**

# LAB 15

## Materials

You may use any of the following materials during your investigation:

**Samples**
- Aluminum (Al)
- Copper (Cu)
- Tin (Sn)
- Zinc (Zn)
- Glass
- Plastic
- Wood

**Consumables**
- Water (in squirt bottle)

**Equipment**
- Graduated cylinder (100 ml)
- 2 Beakers (each 250 ml)
- 2 Polystyrene cups
- Ring clamp and support stand
- Thermometer or temperature probe
- Hot plate
- Electronic or triple beam balance
- Tongs
- Stirring rod
- Safety glasses or goggles
- Chemical-resistant apron
- Nonlatex gloves

## Safety Precautions

Follow all normal lab safety rules. In addition, take the following safety precautions:

1. Wear sanitized indirectly vented chemical-splash goggles and chemical-resistant nonlatex gloves and aprons during lab setup, hands-on activity, and takedown.

2. Use caution when working with hot plates, because they can burn skin and cause fires.

3. Hot plates also need to be kept away from water and other liquids.

4. Use only GFCI-protected electrical receptacles for hot plates.

5. Clean up any spilled liquid immediately to avoid a slip or fall hazard.

6. Handle all glassware with care.

7. Handle glass thermometers with care. They are fragile and can break, causing a sharp hazard that can cut or puncture skin.

8. Wash hands with soap and water after completing the lab activity.

## Investigation Proposal Required?    ☐ Yes        ☐ No

## Getting Started

To calculate the specific heat of a material, you will need to determine how much energy the material is able to transfer to a sample of water using a calorimeter. A calorimeter is used to prevent heat loss to the surroundings (see Figure L15.2). The heat gained by the water in a calorimeter is therefore equal in magnitude (but opposite in sign) to the heat lost by the material:

$$q_{water} = -q_{material}$$

The amount of heat gained by the water is calculated using the mass of water used, the specific heat of water (4.18 J/g•°C), and the difference between the final and initial temperature of the water in the calorimeter. The amount of water used for calorimetry varies, but most people use between 10 and 50 ml because water has such a high specific heat. The equation for calculating the amount of heat gained by the water is

## FIGURE L15.2
**A basic calorimeter**

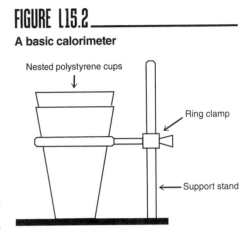

$$q_{water} = m_{water} \times s_{water} \times \Delta T_{water}$$

The amount of heat lost by a material once it is added to the water is calculated using the mass of the material, the specific heat of that material, and the difference between the material's final temperature and its initial temperature. The final temperature of the material is assumed to be the same as the final temperature of the water in the cup. The initial temperature of the material will be 100°C. To ensure that the initial temperature of the material will be 100°C before you add it to the water in the calorimeter, you can place the material in a boiling-water bath for 10–15 minutes. The equation for calculating the amount of heat lost by a metal is

$$-q_{metal} = m_{material} \times s_{material} \times \Delta T_{material}$$

Now that you understand the basics of calorimetry, you must determine what data you need to collect, how you will collect it, and how you will analyze it in order to answer the guiding question.

To determine *what data you will need to collect,* think about the following questions:

- How will you know how much thermal energy has been transferred from a material to the water in a calorimeter?
- What information do you need to calculate the specific heat of material once you know how much thermal energy has been transferred from a material to the water in a calorimeter?

To determine *how you will collect your data,* think about the following questions:

- What equipment will you need to collect the data you need?
- How will you make sure that your data are of high quality (i.e., how will you reduce error)?
- How will you keep track of the data you collect?
- How will you organize your data?

To determine *how you will analyze your data,* think about the following questions:

- What type of calculations will you need to make?
- What type of graph could you create to help make sense of your data?

## Connections to Crosscutting Concepts, the Nature of Science, and the Nature of Scientific Inquiry

As you work through your investigation, be sure to think about

- the importance of defining a system under study;
- how scientists often need track how energy moves into, out of, and within a system;
- the difference between observations and inferences in science; and
- how scientists use different methods to answer different types of questions.

## Initial Argument

Once your group has finished collecting and analyzing your data, your group will need to develop an initial argument. Your initial argument needs to include a *claim, evidence* to support your claim, and a *justification* of the evidence. The claim is your group's answer to the guiding question. The evidence is an analysis and interpretation of your data. Finally, the justification of the evidence is why your group thinks the evidence matters. The justification of the evidence is important because scientists can use different kinds of evidence to support their claims. Your group will create your initial argument on a whiteboard. Your whiteboard should include all the information shown in Figure L15.3.

# FIGURE L15.3 _____

**Argument presentation on a whiteboard**

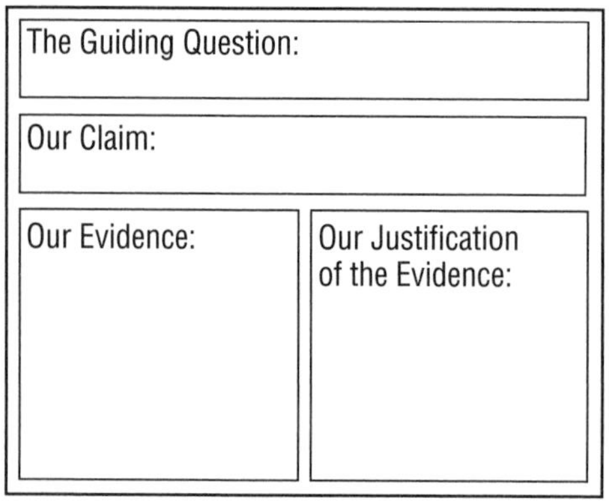

| The Guiding Question: | |
|---|---|
| Our Claim: | |
| Our Evidence: | Our Justification of the Evidence: |

## Argumentation Session

The argumentation session allows all of the groups to share their arguments. One member of each group will stay at the lab station to share that group's argument, while the other members of the group go to the other lab stations to listen to and critique the arguments developed by their classmates. This is similar to how scientists present their arguments to other scientists at conferences. If you are responsible for critiquing your classmates' arguments, your goal is to look for mistakes so these mistakes can be fixed and they can make their argument better. The argumentation session is also a good time to think about ways you can make your initial argument better. Scientists must share and critique arguments like this to develop new ideas.

To critique an argument, you might need more information than what is included on the whiteboard. You will therefore need to ask the presenter lots of questions. Here are some good questions to ask:

- How did you collect your data? Why did you use that method? Why did you collect those data?
- What did you do to make sure the data you collected are reliable? What did you do to decrease measurement error?
- How did your group analyze the data? Why did you decide to do it that way? Did you check your calculations?
- Is that the only way to interpret the results of your analysis? How do you know that your interpretation of your analysis is appropriate?
- Why did your group decide to present your evidence in that way?
- What other claims did your group discuss before you decided on that one? Why did your group abandon those alternative ideas?
- How confident are you that your claim is valid? What could you do to increase your confidence?

Once the argumentation session is complete, you will have a chance to meet with your group and revise your initial argument. Your group might need to gather more data or design a way to test one or more alternative claims as part of this process. Remember, your goal at this stage of the investigation is to develop the most acceptable and valid answer to the research question!

## Report

Once you have completed your research, you will need to prepare an *investigation report* that consists of three sections. Each section should provide an answer to the following questions:

1. What question were you trying to answer and why?
2. What did you do to answer your question and why?
3. What is your argument?

Your report should answer these questions in two pages or less. This report must be typed, and any diagrams, figures, or tables should be embedded into the document. Be sure to write in a persuasive style; you are trying to convince others that your claim is acceptable and valid!

# LAB 15

# Lab 15. Thermal Energy and Specific Heat
## Which Material Has the Greatest Specific Heat?

1. The diagrams below show a 50 g piece of iron and a 50 g piece of tin being added to 50 ml of water in two different calorimeters. The initial temperature of each piece of metal is 100°C. The initial temperature of the water in each calorimeter is 25°C.

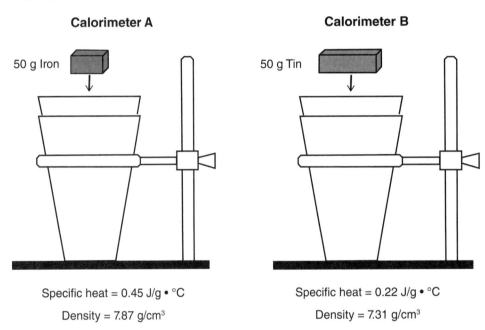

**Calorimeter A**

50 g Iron

Specific heat = 0.45 J/g • °C
Density = 7.87 g/cm³

**Calorimeter B**

50 g Tin

Specific heat = 0.22 J/g • °C
Density = 7.31 g/cm³

What do you think will happen to the temperature of the water in each calorimeter?

a. The temperature of the water in calorimeter A will increase more than it will in calorimeter B.

b. The temperature of the water in calorimeter B will increase more than it will in calorimeter A.

c. The temperature of the water in calorimeters A and B will go up by the same amount.

d. Unsure

How do you know?

2. "Heat from the metal transferred into the water" is an example of an observation.

    a. I agree with this statement.

    b. I disagree with this statement.

Explain your answer, using an example from your investigation about specific heat.

3. Investigations are only scientific if someone designs and then carries out an experiment.

    a. I agree with this statement.

    b. I disagree with this statement.

Explain your answer, using an example from your investigation about specific heat.

4. Scientists often need to define a system before they attempt to study it. Explain what it means to define a system and then explain why it is important in science, using an example from your investigation about specific heat.

5. Scientists often need to track how energy or matter moves into, out of, or within systems to explain a natural phenomenon. Explain why tracking energy or matter is so useful in science, using an example from your investigation about specific heat.

# Lab 16. Electrical Energy and Lightbulbs

## How Does the Arrangement of Lightbulbs That Are Connected to a Battery Affect the Brightness of a Single Bulb in That Circuit?

### Introduction

Scientists and historians generally agree that the approximately 250-year period from the 1550s to about 1800 was one of the most influential periods in history. During this time period—often referred to as the Scientific Revolution—science became increasingly important, and many of the people and developments of this period still influence our society today. For example, scientists such as Copernicus, Galileo, Kepler, and Newton published their most influential works during this time period.

While the ideas of Copernicus, Galileo, Kepler, and Newton (among others) are no doubt important and still remain influential, the most important development from this period may have come from a series of debates between Robert Boyle and Thomas Hobbes. Boyle was an important chemist and inventor. Hobbes was an influential philosopher. Boyle was a member of The Royal Society (along with Isaac Newton, Nicholas Mercator, and Edmond Halley, among many others), a scientific group that began in London during the 1600s (and the oldest scientific group still in existence). Hobbes was not a member of The Royal Society. Boyle and Hobbes had different views on how science should be conducted. Hobbes felt that science should be based on logic and reason, by which he meant that scientists should think about their questions and use philosophical approaches to answer those questions. This, said Hobbes, was how science had been done dating back to Aristotle. Boyle, on the other hand, suggested that science should be based on empirical results (a fancy term for evidence) and scientists should use rigorous investigative methods to answer their questions. Boyle also put forth the idea that scientists need to control for all the potential factors that might affect the outcome of an investigation (a more scientific way of saying that they should account for a factor by keeping it the same across conditions that are being tested). This, said Boyle, was how science should be done in the future, despite how it had been done in the past. After a series of debates and demonstrations, most of the members of The Royal Society sided with Boyle. The reliance on empirical support is what many scientists and historians say is the most important development of the Scientific Revolution.

Another outcome of the Scientific Revolution was the development of new questions that scientists could investigate and attempt to answer. These questions gave rise to entire fields of science, such as microbiology and geology. A third field to develop during this time was the study of electricity—a field that is very much with us today. The study of electricity has led to the development of many important technologies that we still use today. In the year 1800, Alessandro Volta invented the battery. Another important invention due to the study of electricity is the lightbulb, invented by Thomas Edison in 1879.

# Electrical Energy and Lightbulbs

*How Does the Arrangement of Lightbulbs That Are Connected to a Battery Affect the Brightness of a Single Bulb in That Circuit?*

One of the most difficult aspects of inventing a reliable lightbulb was identifying the best material to use for the filament in the lightbulb. The filament is a small wire inside the lightbulb that the electricity must pass through. When the electricity passes through the filament there is a lot of resistance, meaning it is difficult for the electric current to pass through the small wire. As the electric current moves through the filament, it generates heat energy due to the resistance of the wire (similarly to how your hands generate heat energy when you rub them together quickly) and that heat energy causes the wire of the filament to glow. During this process electrical energy is converted to radiant energy (or light energy).

Since the lightbulb and battery were invented, people have been investigating their behavior when they are connected as part of an electric circuit in many different ways. An electric circuit is a continuous path that allows electricity to leave a source (such as a battery), travel through wires and other objects (such as lightbulbs), and then return to the source. Research on batteries, lightbulbs, and circuits has shown that there are two general categories of ways to arrange lightbulbs in an electric circuit and connect them to a battery in a way that will still allow the bulbs to light. The two categories are called series circuits and parallel circuits. When lightbulbs are arranged in series, such as is shown in Figure L16.1(a), each bulb is connected to the next bulb and so forth. When lightbulbs are connected in parallel, such as is shown in Figure L16.1(b), each bulb is connected directly to the battery. The amount of light given off by a lightbulb is influenced in part by the strength of the battery (or other source of electricity) and the ways the lightbulbs are connected together. Scientists have investigated what happens to the brightness of the light emitted by the lightbulb when they are connected in series and parallel. In this investigation, you will have an opportunity to examine how the arrangement of bulbs connected to a battery in an electric circuit affects the brightness of a specific bulb in the circuit.

# FIGURE L16.1

**Bulbs in series (a) and in parallel (b)**

(a)                               (b)

# LAB 16

## Your Task

Use what you know about circuits, the relationship between structure and function, and how to design and carry out an investigation to develop a rule that will allow you to predict the brightness of a bulb based on how it is arranged in an electric circuit. During this investigation, you will want to keep in mind the ideas of Robert Boyle—that scientific rules need empirical support and it is important to control for all the factors that might influence your results during an investigation. Once you develop your rule, you will need to test it to determine if it allows you to predict the brightness of a bulb in a wide range of different circuits.

The guiding question of this investigation is, **How does the arrangement of lightbulbs that are connected to a battery affect the brightness of a single bulb in that circuit?**

## Materials

You may use any of the following materials during your investigation:

- Size D batteries
- Battery holders
- Small lightbulbs
- Lightbulb holders
- Electrical wire
- Light sensor with interface
- Safety glasses or goggles

## Safety Precautions

Follow all normal lab safety rules. In addition, take the following safety precautions:

1. Wear sanitized safety glasses or goggles during lab setup, hands-on activity, and takedown.

2. Use caution when handling bulbs, wires, and batteries. They can get hot and burn skin.

3. Never put batteries in your mouth or on your tongue.

4. Use caution in handling wire ends. They are sharp and can cut or puncture skin.

5. Lightbulbs are made of glass. Be careful handling them. If they break, clean them up immediately and place in a broken glass box.

6. Wash hands with soap and water after completing the lab activity.

## Investigation Proposal Required?   ☐ Yes   ☐ No

## Getting Started

The first step in this investigation is to learn more about how the number and arrangement of bulbs in a circuit affect the brightness of a specific bulb in that circuit. To accomplish this task, you must determine what type of data you need to collect, how you will collect it, and how you will analyze it before you begin.

To determine *what type of data you need to collect,* think about the following questions:

- How will you determine brightness?
- What other factors, besides the type of circuit, could affect the brightness of a lightbulb?
- How will you control for those factors?

To determine *how you will collect your data,* think about the following questions:

- What equipment will you need to collect the data you need?
- How will you make sure that your data are of high quality (i.e., how will you reduce error)?
- How will you keep track of the data you collect?
- How will you organize your data?

To determine *how you will analyze your data,* think about the following questions:

- What factors will you compare to generate your rule?
- What type of table or graph could you create to help make sense of your data?

The second step in this investigation is to develop a rule that you can use to predict the brightness of a bulb in a circuit. Once you have your rule, you will need to test it to determine if it allows you to accurately predict the brightness of a bulb in several new circuits (ones that you did not use to develop your rule). It is important for you to test your rule, because the results of your test will not only allow you to demonstrate that your rule is valid but also will allow you to show that it is a useful way to predict the behavior of a lightbulb when it is connected to a battery and one or more other bulbs. Be sure to modify your rule as needed if it does not allow you to accurately predict the brightness of a bulb in a particular circuit.

## Connections to Crosscutting Concepts, the Nature of Science, and the Nature of Scientific Inquiry

As you work through your investigation, be sure to think about

- the importance of tracking how energy and matter move within electrical systems,
- the relationship between structure and function in nature,
- the difference between data and evidence in science, and
- the nature and role of experiments in science.

# LAB 16

## Initial Argument

Once your group has finished collecting and analyzing your data, your group will need to develop an initial argument. Your initial argument needs to include a *claim, evidence* to support your claim, and a *justification* of the evidence. The claim is your group's answer to the guiding question. The evidence is an analysis and interpretation of your data. Finally, the justification of the evidence is why your group thinks the evidence matters. The justification of the evidence is important because scientists can use different kinds of evidence to support their claims. Your group will create your initial argument on a whiteboard. Your whiteboard should include all the information shown in Figure L16.2.

## FIGURE L16.2
**Argument presentation on a whiteboard**

| The Guiding Question: |  |
| Our Claim: |  |
| Our Evidence: | Our Justification of the Evidence: |

## Argumentation Session

The argumentation session allows all of the groups to share their arguments. One member of each group will stay at the lab station to share that group's argument, while the other members of the group go to the other lab stations to listen to and critique the arguments developed by their classmates. This is similar to how scientists present their arguments to other scientists at conferences. If you are responsible for critiquing your classmates' arguments, your goal is to look for mistakes so these mistakes can be fixed and they can make their argument better. The argumentation session is also a good time to think about ways you can make your initial argument better. Scientists must share and critique arguments like this to develop new ideas.

To critique an argument, you might need more information than what is included on the whiteboard. You will therefore need to ask the presenter lots of questions. Here are some good questions to ask:

- How did you collect your data? Why did you use that method? Why did you collect those data?

- What did you do to make sure the data you collected are reliable? What did you do to decrease measurement error?

- How did your group analyze the data? Why did you decide to do it that way? Did you check your calculations?

- Is that the only way to interpret the results of your analysis? How do you know that your interpretation of your analysis is appropriate?

- Why did your group decide to present your evidence in that way?

- What other claims did your group discuss before you decided on that one? Why did your group abandon those alternative ideas?

National Science Teachers Association

- How confident are you that your claim is valid? What could you do to increase your confidence?

Once the argumentation session is complete, you will have a chance to meet with your group and revise your initial argument. Your group might need to gather more data or design a way to test one or more alternative claims as part of this process. Remember, your goal at this stage of the investigation is to develop the most acceptable and valid answer to the research question!

## Report

Once you have completed your research, you will need to prepare an *investigation report* that consists of three sections. Each section should provide an answer to the following questions:

- What question were you trying to answer and why?
- What did you do to answer your question and why?
- What is your argument?

Your report should answer these questions in two pages or less. This report must be typed, and any diagrams, figures, or tables should be embedded into the document. Be sure to write in a persuasive style; you are trying to convince others that your claim is acceptable and valid!

## Checkout Questions

# Lab 16. Electrical Energy and Lightbulbs

## How Does the Arrangement of Lightbulbs That Are Connected to a Battery Affect the Brightness of a Single Bulb in That Circuit?

1. In which circuit will lightbulb A be brightest: circuit 1, circuit 2, or circuit 3?

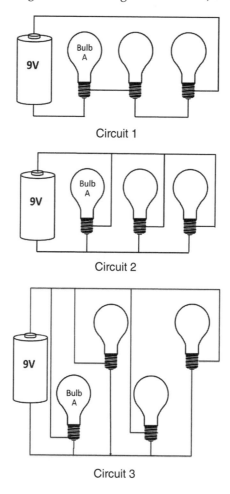

Circuit 1

Circuit 2

Circuit 3

Explain your answer. Why do you think that bulb will be brightest?

## Electrical Energy and Lightbulbs

*How Does the Arrangement of Lightbulbs That Are Connected to a Battery Affect the Brightness of a Single Bulb in That Circuit?*

2. If you want to make a lightbulb less bright, what are some things you could do to the circuit to make the bulb decrease in brightness? How do you know those things will work?

3. In science, there is no difference between data and evidence.

   a. I agree with this statement.

   b. I disagree with this statement.

   Explain your answer, using an example from your investigation about series and parallel circuits.

4. No matter what is being investigated, conducting an experiment is the best way to develop scientific knowledge.

   a. I agree with this statement.

   b. I disagree with this statement.

Explain your answer, using an example from your investigation about series and parallel circuits.

5. Often, the structure of a system influences the function of that system. In other words, how something is structured influences how it works. Using an example from your investigation about series and parallel circuits, explain how the structure of a system can influence how it functions.

6. Scientists often need to keep track of the movement of energy into, out of, and within systems. Using an example from your investigation about series and parallel circuits, explain why it is important to track how input of energy into a system affects how it behaves.

# Application Labs

## Lab Handout

# Lab 17. Rate of Energy Transfer
## How Does the Surface Area of a Substance Affect the Rate at Which Thermal Energy Is Transferred From One Substance to Another?

### Introduction

Understanding how energy is transferred from one object or substance to another is an important concept within science. Many common events involve a transfer of energy, such as heating a pot of water on the stove or when a car engine burns gasoline. In the first example, heat energy is transferred from the stove to the pot of water, and the water absorbs the heat energy and will eventually begin to boil. In a car, the chemical energy stored in the gasoline is released when it is burned in the engine; that chemical energy is ultimately converted to kinetic energy and results in the motion of the car.

The law of conservation of energy indicates that energy is not created or destroyed, only converted from one form to another. There are many different types of energy that can be transferred between objects. When two objects are at different temperatures, it is possible for heat or thermal energy to transfer from one object to the other. If you place a cold pot of water on a hot stove burner, for example, thermal energy will transfer from the stove to the water and the water will get warmer. Heat energy always moves from objects with a high temperature toward objects with a lower temperature.

When we measure the temperature of an object, we often use the Celsius scale. On the Celsius scale water freezes at 0°C and boils at 100°C. The temperature of a substance is a measure of the average kinetic energy of the particles of that substance. Water molecules in a cold sample of water at 10°C have less kinetic energy than water molecules in a sample of hot water at 50°C. In this example, the water molecules at 50°C will be moving faster. If these two samples of water with different temperatures are mixed together, the fast- and slow-moving particles will transfer energy until eventually the molecules all have similar amounts of kinetic energy. When molecules with higher kinetic energy bump into molecules with lower kinetic energy, the faster-moving particles transfer kinetic energy to the slower particles. The transfer of energy will result in the water mixture having a temperature that is in the middle of the starting temperatures, which is called an equilibrium temperature—in this example, perhaps about 30°C.

Whenever substances at different temperatures come into contact with one another, thermal energy will be transferred from the hotter object to the cooler object until an equilibrium temperature is reached. The rate at which that thermal energy is transferred from one substance to another, however, is based on several factors. Some of those factors include the properties of the specific substances, the amount of the substances involved, the starting temperatures, and the size and shape of the objects (see Figure L17.1 on page 306 for three samples of metal with equal mass but different surface areas). In this activity,

## Rate of Energy Transfer

*How Does the Surface Area of a Substance Affect the Rate at Which Thermal Energy Is Transferred From One Substance to Another?*

you will investigate how surface area affects the rate of heat transfer.

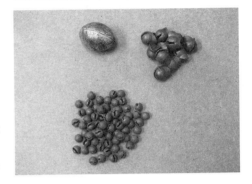

**FIGURE L17.1**

**Three samples of metal with equal mass but different surface areas**

## Your Task

Use what you know about thermal energy, tracking energy, and the relationship between structure and function to design and carry out an investigation that will allow you to test how the surface area of a hot object affects the rate at which thermal energy is transferred from that object to water. To complete this task, you will need to heat up several objects with different surface areas and then place them into room-temperature water. It is up to your group to determine how much and at what rate thermal energy is transferred to the room-temperature water.

The guiding question of this investigation is, **How does the surface area of a substance affect the rate at which thermal energy is transferred from one substance to another?**

## Materials

You may use any of the following materials during your investigation:

**Consumable**
- Water

**Equipment**
- Metal samples
- Hot plate
- Beaker (1,000 ml)
- Graduated cylinder (100 ml)
- Tongs
- Mesh bags
- Styrofoam cups
- Electronic or triple beam balance
- Thermometer or temperature probe
- Ruler
- Stopwatch
- Safety glasses or goggles
- Chemical-resistant apron
- Nonlatex gloves

## Safety Precautions

Follow all normal lab safety rules. In addition, take the following safety precautions:

1. Wear sanitized indirectly vented chemical-splash goggles and chemical-resistant nonlatex gloves and aprons during lab setup, hands-on activity, and takedown.

2. Use caution when working with hot plates, because they can burn skin and cause fires.

3. Hot plates also need to be kept away from water and other liquids.

4. Use caution when working with hot water, because it can burn skin.

5. Only use GFCI-protected electrical receptacles for hot plates.

6. Clean up any spilled liquid immediately to avoid a slip or fall hazard.

7. Never put consumables in your mouth.

8. Always use tongs to move the heated metal.

9. Handle all glassware with care.

10. Handle glass thermometers with care. They are fragile and can break, causing a sharp hazard that can cut or puncture skin.

11. Never return the consumables to stock bottles.

12. Wash hands with soap and water after completing the lab activity.

## Investigation Proposal Required?    ☐ Yes        ☐ No

## Getting Started

To answer the guiding question, you will need to design and conduct an investigation to measure the rate at which thermal energy is transferred to the water. To accomplish this task, you must determine what type of data you need to collect, how you will collect it, and how you will analyze it.

To determine *what type of data you need to collect*, think about the following questions:

- How will you determine the amount of energy transferred?
- What information or measurements will you need to record?
- How will you know when the equilibrium temperature is achieved?
- How will you measure the surface area of the different samples?
- What variables will you control from one sample to the next?

To determine *how you will collect your data*, think about the following questions:

- What equipment will you need to collect the data you need?
- How will you make sure that your data are of high quality (i.e., how will you reduce error)?
- Are there different ways you can measure the amount of energy transferred?
- How will you keep track of the data you collect?
- How will you organize your data?

# Rate of Energy Transfer

*How Does the Surface Area of a Substance Affect the Rate at Which Thermal Energy Is Transferred From One Substance to Another?*

To determine *how you will analyze your data,* think about the following questions:

- How will you determine the rate of heat transfer?
- What type of table or graph could you create to help make sense of your data?

## Connections to Crosscutting Concepts, the Nature of Science, and the Nature of Scientific Inquiry

As you work through your investigation, be sure to think about

- why it is important to track how energy and matter move into, out of, and within systems;
- how the structure or shape of something can influence how it functions and places limits on what it can and cannot do;
- the difference between observations and inferences in science; and
- the different methods used in science.

## Initial Argument

Once your group has finished collecting and analyzing your data, your group will need to develop an initial argument. Your initial argument needs to include a *claim, evidence* to support your claim, and a *justification* of the evidence. The claim is your group's answer to the guiding question. The evidence is an analysis and interpretation of your data. Finally, the justification of the evidence is why your group thinks the evidence matters. The justification of the evidence is important because scientists can use different kinds of evidence to support their claims. Your group will create your initial argument on a whiteboard. Your whiteboard should include all the information shown in Figure L17.2.

## FIGURE L17.2

**Argument presentation on a whiteboard**

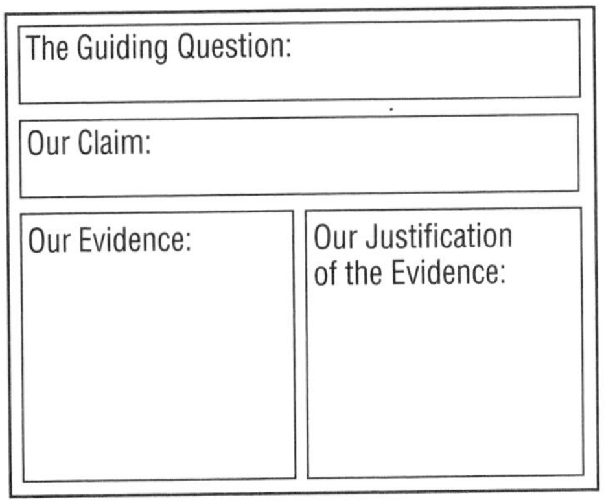

| The Guiding Question: | |
|---|---|
| Our Claim: | |
| Our Evidence: | Our Justification of the Evidence: |

## Argumentation Session

The argumentation session allows all of the groups to share their arguments. One member of each group will stay at the lab station to share that group's argument, while the other members of the group go to the other lab stations to listen to and critique the arguments developed by their classmates. This is similar to how scientists present their arguments to other scientists at conferences. If you are responsible for critiquing your classmates' arguments, your goal is to look for mistakes so these mistakes can be fixed and they can make their argument better. The argumentation session is also a good time to think about ways you can make your initial argument better. Scientists must share and critique arguments like this to develop new ideas.

To critique an argument, you might need more information than what is included on the whiteboard. You will therefore need to ask the presenter lots of questions. Here are some good questions to ask:

- How did you collect your data? Why did you use that method? Why did you collect those data?

- What did you do to make sure the data you collected are reliable? What did you do to decrease measurement error?

- How did your group analyze the data? Why did you decide to do it that way? Did you check your calculations?

- Is that the only way to interpret the results of your analysis? How do you know that your interpretation of your analysis is appropriate?

- Why did your group decide to present your evidence in that way?

- What other claims did your group discuss before you decided on that one? Why did your group abandon those alternative ideas?

- How confident are you that your claim is valid? What could you do to increase your confidence?

Once the argumentation session is complete, you will have a chance to meet with your group and revise your initial argument. Your group might need to gather more data or design a way to test one or more alternative claims as part of this process. Remember, your goal at this stage of the investigation is to develop the most acceptable and valid answer to the research question!

## Report

Once you have completed your research, you will need to prepare an *investigation report* that consists of three sections. Each section should provide an answer to the following questions:

1. What question were you trying to answer and why?

2. What did you do to answer your question and why?

3. What is your argument?

Your report should answer these questions in two pages or less. This report must be typed, and any diagrams, figures, or tables should be embedded into the document. Be sure to write in a persuasive style; you are trying to convince others that your claim is acceptable and valid!

## *Checkout Questions*

# Lab 17. Rate of Energy Transfer
## How Does the Surface Area of a Substance Affect the Rate at Which Thermal Energy Is Transferred From One Substance to Another?

1. Denise was conducting an investigation on how long it would take ice to melt. She decided to test ice cubes versus crushed ice. She put 500 ml of 25°C water into two cups, and then she put 200 g of ice into each cup. One cup had ice cubes and the other cup had small crushed ice. She recorded the time it took for the ice in each cup to melt; the setup and results are shown below.

200 g cubed ice before melting

Time to melt: 14 minutes

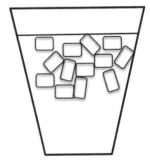

200 g crushed ice before melting

Time to melt: 9 minutes

Use what you know about energy transfer to explain the results that Denise obtained.

2. An engineer needs to put a hot piece of metal and a cold piece of metal together in a way that makes them reach their equilibrium temperature the fastest. Below are four options that she has come up with; in each option, the gray bar is hot and the black bar is cold. Which option would you recommend?

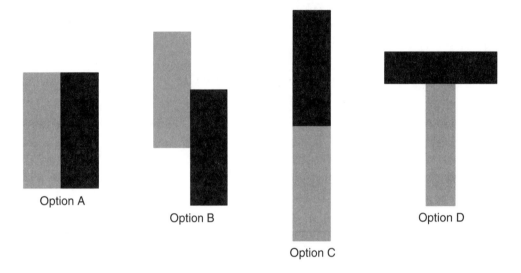

Option A

Option B

Option C

Option D

Explain why you chose that option to recommend to the engineer.

3. It is more important for scientists to make observations than inferences.

   a. I agree with this statement.

   b. I disagree with this statement.

   Explain your answer, using an example from your investigation on energy transfer.

# Rate of Energy Transfer

*How Does the Surface Area of a Substance Affect the Rate at Which Thermal Energy Is Transferred From One Substance to Another?*

4. Different scientists may use different procedures to investigate the same question.

   a. I agree with this statement.

   b. I disagree with this statement.

   Explain your answer, using an example from your investigation on energy transfer.

5. Understanding how systems work is an important aspect of science and engineering. Use an example from your investigation about energy transfer to help explain why it is important to track how energy and matter move into, out of, and within systems.

6. Scientists often study the structure of objects because the structure can provide useful clues about the function of that object. Explain why it is important for scientists to understand the connection between structure and function, using an example from your investigation on energy transfer.

## Lab Handout

# Lab 18. Radiation and Energy Transfer
## What Color Should We Paint a Building to Reduce Cooling Costs?

### Introduction

Radiant energy is the energy transported by electromagnetic waves. Electromagnetic waves transport many different types of energy (see Figure L18.1). The microwaves that warm up your food when you place it into a microwave oven are electromagnetic waves, as are the x-rays that a doctor or dentist uses to take pictures of your bones. In fact, everything you see is also due to electromagnetic waves. Visible light, the light that humans can see, travels in waves. Each color has its own wavelength, which corresponds to a different amount of energy. When those waves reach our eyes, they can then be processed and perceived as color. Certain properties of an object cause it to reflect one wavelength of light and absorb others. For example, the reason an object appears blue is because it absorbs all other wavelengths and reflects a blue wavelength of light, which your eyes receive and, along with your brain, process as the color blue.

# FIGURE L18.1

**The electromagnetic spectrum**

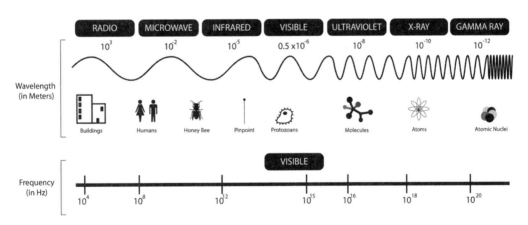

As is true for all energy, radiant energy can be transferred into other forms but cannot be created or destroyed. However, radiant energy is different in that it does not need a medium (matter), such as air or metal, to travel. Radiant energy can travel through a vacuum, such as space. The Sun emits radiant energy that travels through space. Some of that energy reaches Earth. When radiant energy reaches an object, it increases the rate of vibration of the atoms and/or molecules in that object, raising its overall temperature.

# LAB 18

Radiant energy from the Sun raises the temperature of nearly everything on Earth, but some things are more affected than others.

When a new building is designed, its architects take into account the future energy costs of the building. Energy-efficient buildings are cheaper and more efficient to own and operate, and are also better for the environment. Some energy-saving or energy-storing measures, such as advanced heating and cooling systems, are expensive, but other measures, such as insulation or paint color, are simpler and less expensive. However, each step taken to increase the energy efficiency of a building is beneficial, not only for those who will own and use the building but for everyone, because we all benefit from the reduced use of energy resources.

## Your Task

Use what you know about electromagnetic waves, visible light, and energy transfer to design and conduct an experiment to determine which paint color keeps a building the coolest, reducing its cooling costs.

The guiding question of this investigation is, **What color should we paint a building to reduce cooling costs?**

## Materials

You may use any of the following materials during your investigation:

- Canisters with various paint colors
- Heat lamp
- Stopwatch
- Thermometer or temperature probe
- Safety glasses or goggles

## Safety Precautions

Follow all normal lab safety rules. In addition, take the following safety precautions:

- Wear sanitized safety glasses or goggles during lab setup, hands-on activity, and takedown.
- Use caution when working with heat lamps and metal containers, because they get hot and can burn skin.
- Clamp lamps to a secure structure out of the way of foot traffic, and avoid touching parts of the lamp other than the power switch until it has cooled.
- Only use GFCI-protected electrical receptacles for the heat lamp.
- Handle glass thermometers with care. They are fragile and can break, causing a sharp hazard that can cut or puncture skin.
- Wash hands with soap and water after completing the lab activity.

**Investigation Proposal Required?**  ☐ Yes  ☐ No

## Getting Started

To answer the guiding question, you will need to design an experiment that will allow you to determine which exterior paint color is associated with the lowest average canister temperature. To accomplish this task, you can heat canisters that are painted different colors using a heat lamp (see Figure L18.2). Before you can begin heating different canisters, you must first determine what type of data you need to collect, how you will collect it, and how you will analyze it.

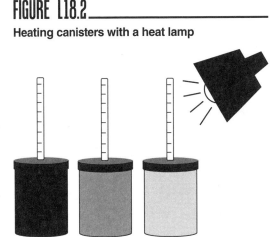

## FIGURE L18.2

**Heating canisters with a heat lamp**

To determine *what type of data you need to collect,* think about the following questions:

- What will serve as your independent variable in the investigation?
- What will serve as your dependent variable in the investigation?
- What types of measurements will you need to make?

To determine *how you will collect your data,* think about the following questions:

- What equipment will you use to take measurements?
- When will you make the measurements that you need?
- What other factors will you need to control during your experiment?
- How will you make sure that your data are of high quality (i.e., how will you reduce error)?
- How will you keep track of the data you collect?
- How will you organize your data?

To determine *how you will analyze your data,* think about the following questions:

- What type of calculations will you need to make?
- What type of table or graph could you create to help make sense of your data?

## Connections to Crosscutting Concepts, the Nature of Science, and the Nature of Scientific Inquiry

As you work through your investigation, be sure to think about

- how scientists work to explain the relationships between causes and effects;

- the importance of tracking how energy moves into, out of, and within systems;
- the difference between scientific laws and scientific theories; and
- the nature and role of experiments in science.

## Initial Argument

Once your group has finished collecting and analyzing your data, your group will need to develop an initial argument. Your initial argument needs to include a *claim, evidence* to support your claim, and a *justification* of the evidence. The claim is your group's answer to the guiding question. The evidence is an analysis and interpretation of your data. Finally, the justification of the evidence is why your group thinks the evidence matters. The justification of the evidence is important because scientists can use different kinds of evidence to support their claims. Your group will create your initial argument on a whiteboard. Your whiteboard should include all the information shown in Figure L18.3.

## FIGURE L18.3 _____

**Argument presentation on a whiteboard**

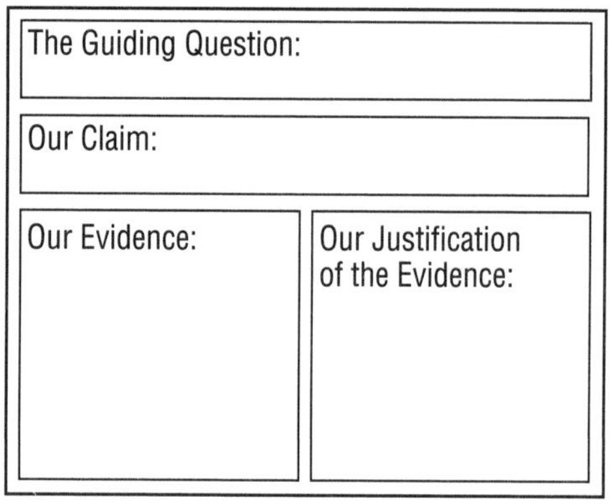

| The Guiding Question: | |
|---|---|
| Our Claim: | |
| Our Evidence: | Our Justification of the Evidence: |

## Argumentation Session

The argumentation session allows all of the groups to share their arguments. One member of each group will stay at the lab station to share that group's argument, while the other members of the group go to the other lab stations to listen to and critique the arguments developed by their classmates. This is similar to how scientists present their arguments to other scientists at conferences. If you are responsible for critiquing your classmates' arguments, your goal is to look for mistakes so these mistakes can be fixed and they can make their argument better. The argumentation session is also a good time to think about ways you can make your initial argument better. Scientists must share and critique arguments like this to develop new ideas.

To critique an argument, you might need more information than what is included on the whiteboard. You will therefore need to ask the presenter lots of questions. Here are some good questions to ask:

- How did you collect your data? Why did you use that method? Why did you collect those data?
- What did you do to make sure the data you collected are reliable? What did you do to decrease measurement error?
- How did your group analyze the data? Why did you decide to do it that way? Did you check your calculations?

- Is that the only way to interpret the results of your analysis? How do you know that your interpretation of your analysis is appropriate?

- Why did your group decide to present your evidence in that way?

- What other claims did your group discuss before you decided on that one? Why did your group abandon those alternative ideas?

- How confident are you that your claim is valid? What could you do to increase your confidence?

Once the argumentation session is complete, you will have a chance to meet with your group and revise your initial argument. Your group might need to gather more data or design a way to test one or more alternative claims as part of this process. Remember, your goal at this stage of the investigation is to develop the most acceptable and valid answer to the research question!

## Report

Once you have completed your research, you will need to prepare an *investigation report* that consists of three sections. Each section should provide an answer to the following questions:

1. What question were you trying to answer and why?

2. What did you do to answer your question and why?

3. What is your argument?

Your report should answer these questions in two pages or less. This report must be typed, and any diagrams, figures, or tables should be embedded into the document. Be sure to write in a persuasive style; you are trying to convince others that your claim is acceptable and valid!

# LAB 18

## Lab 18. Radiation and Energy Transfer
### What Color Should We Paint a Building to Reduce Cooling Costs?

1. Each of the objects below was left in sunlight for one hour. Each object reflects a different wavelength of light, as shown below, and absorbs the other wavelengths. At the end of the hour, which object would you expect to have the highest average temperature?

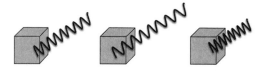

   Explain your answer. Why do you think that object will have the highest average temperature at the end of the hour?

2. Shelby is choosing a solar panel to gather energy from the Sun for heating her home. There is one model available, but it comes in three colors. The first color reflects light with wavelengths between 350 and 400 nm, the second between 600 and 700 nm, and the third reflects no visible light. Which option is the best choice for Shelby if she wants her solar panel to absorb the most energy possible?

   Explain your answer. Why did you recommend that option to Shelby?

3. In science, laws are more important than theories.

   a. I agree with this statement.

   b. I disagree with this statement.

   Explain your answer, using an example from your investigation about radiant energy transfer.

4. Regardless of the question you want to answer, an experiment is always the best way to conduct a scientific investigation.

   a. I agree with this statement.

   b. I disagree with this statement.

   Explain your answer, using an example from your investigation about radiant energy transfer.

5. Scientists often need to look for and understand the underlying cause of events in nature. Explain why it is important to be able to identify and understand cause-and-effect relationships in science, using an example from your investigation about radiant energy transfer.

6. Scientists often need to keep track of the flow of energy within systems. Using an example from your investigation about radiant energy transfer, explain why it is important to keep track of how energy moves into, out of, and within systems.

# SECTION 5
# Physical Science
# Core Idea 4

# Waves and Their Applications in Technologies for Information Transfer

# Introduction Labs

# Lab 19. Wave Properties

## How Do Frequency, Amplitude, and Wavelength of a Transverse Wave Affect Its Energy?

### Introduction

Energy can be transported by waves. There are many forms of waves that exist in the world. Mechanical waves, such as sound waves or water waves, must travel through a medium, or matter. For example, when you speak, you create a pressure disturbance in the air that travels as a wave through the air molecules. You can also create a wave in a rope or string by moving one end from side to side. In each case, the wave travels through the medium, the air or the rope (or string). Electromagnetic waves, such as radio, ultraviolet, and visible light waves, don't require a medium to travel. Instead, the vibrations of perpendicular electric and magnetic fields form these waves.

Although waves may travel differently, some mechanical and electromagnetic waves can be represented by the same basic shape, or waveform. Electromagnetic waves and the waves you might make in a rope or string, for example, are called transverse waves. A drawing of a transverse wave is shown in Figure L19.1. The highest point of the wave is called the crest, and the lowest point is called the trough. The wavelength of the wave, a measure of how long the wave is, can be found by measuring the distance between the same point on a wave and the wave in front of or behind it. Usually, this is done by measuring crest to crest or trough to trough. The amplitude of a wave is the distance from the resting position (the horizontal line) to the crest or trough. The frequency of a wave is hard to show in a picture. A wave's frequency is a measure of how many times a wave passes a certain point in a certain amount of time. To measure frequency, scientists measure the number of wave cycles (trough to trough or crest to crest) that occur in 1 second, and they measure this value in hertz (Hz). One cycle per second is 1 Hz, two per second is 2 Hz, and so on.

## FIGURE L19.1 _____

**Transverse wave**

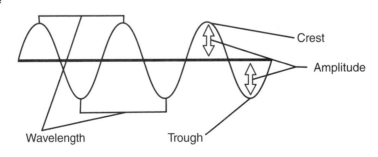

The properties of a wave contain information about the energy that wave is carrying and also determine its use. For example, electromagnetic radio waves are used to transmit the radio signals your car stereo picks up. Your favorite station numbers are actually measurements of the frequency at which that station broadcasts.

## Your Task

Use what you know about waves and energy to design and carry out an investigation that will allow you to describe the relationship between a wave's energy and its amplitude, wavelength, and frequency.

The guiding question of this investigation is, **How do frequency, amplitude, and wavelength of a transverse wave affect its energy?**

## Materials

You will use an online simulation called *Wave on a String* to conduct your investigation. You can access the simulation by going to the following website: *https://phet.colorado.edu/sims/html/wave-on-a-string/latest/wave-on-a-string_en.html*.

## Safety Precautions

Follow all normal lab safety rules.

## Investigation Proposal Required?     ☐ Yes        ☐ No

## Getting Started

To answer the guiding question, you will need to design and carry out an experiment. To accomplish this task, you must determine what type of data you need to collect, how you will collect it, and how you will analyze it. The *Wave on a String* simulation allows you to propagate (start) and manipulate a wave on a virtual rope. The rope is shown as a series of red circles, with every ninth circle colored green (see Figure L19.2). This will make it easier for you to track and measure the properties of the various waves you create.

The upper-left-hand corner of the screen has a box with options for manual, oscillate, and pulse. These options allow you to choose how you will make the waves you will use for data. The "Manual" option requires that you move a wrench up and down to create a wave. The "Oscillate" option creates the wave for you, and you can adjust the frequency and amplitude of the waves using a slider that will appear at the bottom of the screen. Do not choose the "Pulse" option. Because this option only moves upward, it does not create a transverse wave and will not produce a wave that will be helpful for your investigation. You also have the option to use a rope with a fixed end, a loose end, or no end. The simulation provides rulers, a timer, and a reference line for you to use. To activate these tools, simply check the box next to each option. You can move the rulers by clicking

**A screenshot of the *Wave on a String* simulation**

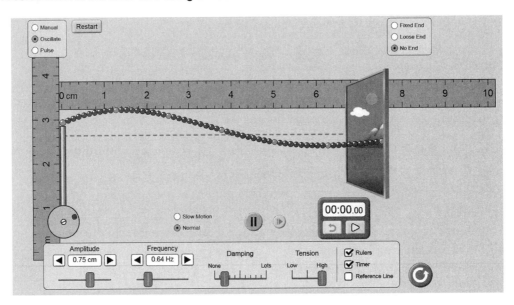

and dragging them to different locations. You may start and pause the simulation at any time by selecting the play/pause button at the bottom of the screen. You can also view the simulation in normal time or in slow motion.

You will need to design and carry out at least three different experiments using the *Wave on a String* simulation in order to determine the relationship between frequency, amplitude, wavelength, and energy. You will need to conduct at least three different experiments, because you will need to be able to answer three specific questions before you will be able to develop an answer to the guiding question:

- How does changing the frequency affect the energy of the wave?
- How does changing the amplitude affect the energy of the wave?
- How does changing the wavelength affect the energy of the wave?

It will be important for you to determine what type of data you need to collect, how to collect the data you need, and how you will need to analyze your data for each experiment, because each experiment is slightly different.

To determine *what type of data you need to collect,* think about the following questions:

- What will serve as your independent variable in the investigation?
- What will serve as your dependent variable(s) in the investigation?
- How will you define and determine the amount of energy being put into the waves?

- How will you measure the various properties of the waves?

To determine *how you will collect your data,* think about the following questions:

- What simulation settings will you use to collect the data you need?
- How will you make sure that your data are of high quality (i.e., how will you reduce error)?
- How will you keep track of the data you collect?
- How will you organize your data?

To determine *how you will analyze your data,* think about the following questions:

- What type of calculations will you need to make?
- What type of table or graph could you create to help make sense of your data?
- How will you determine if there is a relationship between different variables?

## Connections to Crosscutting Concepts, the Nature of Science, and the Nature of Scientific Inquiry

As you work through your investigation, be sure to think about

- how scientists work to explain the relationships between causes and effects;
- the importance of tracking how energy moves into, out of, and within systems;
- the difference between data and evidence in science; and
- methods used in scientific investigations.

## Initial Argument

Once your group has finished collecting and analyzing your data, your group will need to develop an initial argument. Your initial argument needs to include a *claim, evidence* to support your claim, and a *justification* of the evidence. The claim is your group's answer to the guiding question. The evidence is an analysis and interpretation of your data. Finally, the justification of the evidence is why your group thinks the evidence matters. The justification of the evidence is important because scientists can use different kinds of evidence to support their claims. Your group will create your initial argument on a whiteboard. Your whiteboard should include all the information shown in Figure L19.3.

## Argumentation Session

The argumentation session allows all of the groups to share their arguments. One member of each group will stay at the lab station to share that group's argument, while the other members of the group go to the other lab stations to listen to and critique the arguments developed by their classmates. This is similar to how scientists present their arguments

# LAB 19

## FIGURE L19.3 _____

**Argument presentation on a whiteboard**

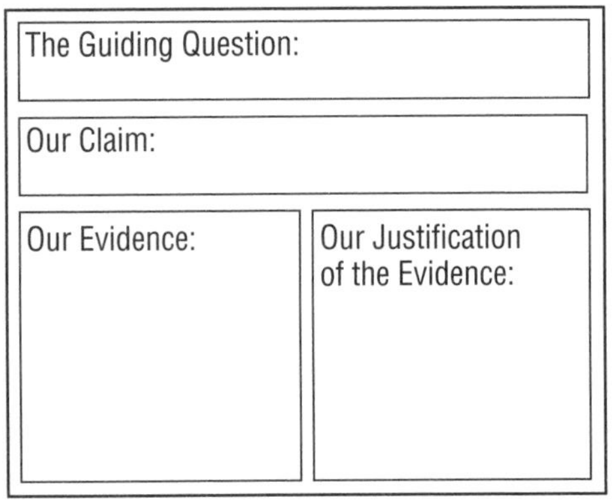

| The Guiding Question: | |
|---|---|
| Our Claim: | |
| Our Evidence: | Our Justification of the Evidence: |

to other scientists at conferences. If you are responsible for critiquing your classmates' arguments, your goal is to look for mistakes so these mistakes can be fixed and they can make their argument better. The argumentation session is also a good time to think about ways you can make your initial argument better. Scientists must share and critique arguments like this to develop new ideas.

To critique an argument, you might need more information than what is included on the whiteboard. You will therefore need to ask the presenter lots of questions. Here are some good questions to ask:

- How did you collect your data? Why did you use that method? Why did you collect those data?

- What did you do to make sure the data you collected are reliable? What did you do to decrease measurement error?

- How did your group analyze the data? Why did you decide to do it that way? Did you check your calculations?

- Is that the only way to interpret the results of your analysis? How do you know that your interpretation of your analysis is appropriate?

- Why did your group decide to present your evidence in that way?

- What other claims did your group discuss before you decided on that one? Why did your group abandon those alternative ideas?

- How confident are you that your claim is valid? What could you do to increase your confidence?

Once the argumentation session is complete, you will have a chance to meet with your group and revise your initial argument. Your group might need to gather more data or design a way to test one or more alternative claims as part of this process. Remember, your goal at this stage of the investigation is to develop the most acceptable and valid answer to the research question!

## Report

Once you have completed your research, you will need to prepare an *investigation report* that consists of three sections. Each section should provide an answer to the following questions:

1. What question were you trying to answer and why?

2. What did you do to answer your question and why?

3. What is your argument?

Your report should answer these questions in two pages or less. This report must be typed, and any diagrams, figures, or tables should be embedded into the document. Be sure to write in a persuasive style; you are trying to convince others that your claim is acceptable and valid!

## Checkout Questions

# Lab 19. Wave Properties
## How Do Frequency, Amplitude, and Wavelength of a Transverse Wave Affect Its Energy?

1. Order the transverse waves below from greatest to least energy carried:

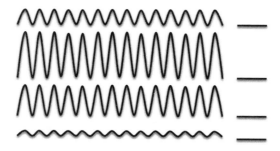

Explain your answer. Why do you think the order that you chose is correct?

2. Jalen moves the end of a rope to produce the wave shown below:

Without changing how far up or down he moves his arm, Jalen moves the end of the rope faster than before, working harder to move the rope. Which wave shown below looks like the wave Jalen is now making?

Original Wave:

Option 1:

Option 2:

Option 3:

Explain your answer. Why do you think the wave Jalen is making looks like the option you chose?

3. In science, there is no difference between data and evidence.

   a. I agree with this statement.

   b. I disagree with this statement.

   Explain your answer, using an example from your investigation about the properties of transverse waves.

4. No matter what is being investigated, conducting an experiment is the best way to develop scientific knowledge.

   a. I agree with this statement.

   b. I disagree with this statement.

   Explain your answer, using an example from your investigation about the properties of transverse waves.

5. Often, changing one part of a system will cause another part of that system to change as well. Determining the cause of observed effects is an important pursuit in science. Using an example from your investigation about properties of transverse waves, explain why it is helpful for scientists to investigate cause-and-effect relationships in the natural world.

6. Scientists often need to keep track of the movement of energy into, out of, and within systems. Using an example from your investigation about properties of transverse waves, explain why it is important to track how input of energy into a system affects how it behaves.

## Lab Handout

# Lab 20. Reflection and Refraction

## How Can You Predict Where a Ray of Light Will Go When It Comes in Contact With Different Types of Transparent Materials?

### Introduction

Our understanding of the nature of light and how it behaves has changed a great deal over the centuries. The first real explanations for the nature and behavior of light came from the ancient Greeks. Most of these early models describe the nature of light as a ray. A ray moves in a straight line from one point to another. Euclid and Ptolemy, for example, used ray diagrams to show how light bounces off a smooth surface or bends as it passes from one transparent medium to another. Other scholars took these ideas and refined them to explain the behavior of light when it strikes a mirror, a lens, or a prism. This field of study is now called geometrical optics. The most famous practitioner of geometrical optics was the 10th–11th century Arab scientist Ibn al-Haytham, who developed mathematical equations that describe how light bends as it travels through different media.

Scientists began to use different models to explain the nature of light in the 17th century. For example, Christiaan Huygens claimed that light is a wave that moves through an "invisible ether" that exists all around us. Isaac Newton, in contrast, claimed that light is composed of small particles, because it travels in a straight line and bounces off a mirror, much like a ball bounces off a wall. Most scientists continued to use a model that treated light as particle in their research until the early part of the 19th century. In 1801, however, Thomas Young showed that if light is made to travel through two slits in a card, it produces a series of light and dark bands on a screen. He argued that this observation would not be possible if light was composed of particles that travel in a straight line (see Figure L20.1[a]) but it would be possible if it traveled through space and time as a wave (see Figure L20.1[b]).

Then in the 1860s, James Maxwell created a new model that described the nature of light as electromagnetic radiation. Electromagnetic radiation does not need a medium to travel through like sound waves do, and when it is traveling in a vacuum (such as space), it moves at a speed of about 300,000 kilometers per second. According to this model, light waves come in many different sizes and these waves can be described in terms of wavelength and

## FIGURE L20.1

**Appearance of light on screen if light is composed of particles (a) or waves (b)**

(a)

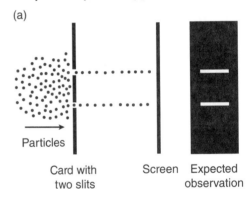

Particles

Card with two slits     Screen    Expected observation

(b)

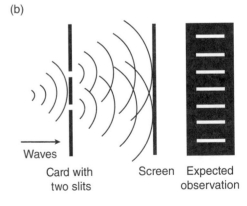

Waves

Card with two slits     Screen    Expected observation

frequency (see Figure L20.2). The wavelengths of light that we can see are between 400 and 700 nanometers long, but all the different wavelengths in the electromagnetic spectrum range from 0.1 nanometer (gamma rays) to several meters (radio waves) in length. The frequency of a light wave is the number of waves that pass a point in space in a specific time interval. We measure frequency in hertz (cycles per second), abbreviated Hz. Red light has a frequency of 430 trillion Hz, and violet light has a frequency of 750 trillion Hz.

## FIGURE L20.2

**Wavelengths and frequencies of the different types of waves in the electromagnetic spectrum**

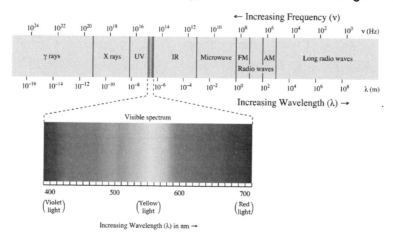

As it turns out, all of these models for the nature of light are both right and wrong at the same time, because they can only be used to explain or predict certain behaviors of light. Scientists now use a model that describes the nature of light as being a particle and a wave. In this investigation, however, you will use a ray model of light to investigate how light behaves when it comes in contact with different types of transparent materials. When a ray of light passes between two transparent materials (such as air, water, plastic, or glass), part of the ray is reflected and stays in the first material, while the rest of the ray is refracted as it passes into the second material. The ray of light refracts when it enters the second material because it changes speed (slows down or speeds up) as it begins to travel through the new materials.

Figure L20.3 shows a ray of light crossing the boundary between two transparent materials. In the field of optics, a line perpendicular to the boundary is used to measure the angles of the light rays. This line is called the surface normal. The angle the incoming ray makes with the surface normal is called the angle of incidence ($\theta_i$). The angle the reflected ray makes with the normal is called the angle of reflection ($\theta_r$), and the angle the refracted ray makes with the normal is called the angle of refraction ($\theta_R$). Your goal in this investigation is to develop one or more rules that you can use to predict the behavior and path of the reflected and refracted rays, much like Ibn al-Haytham did when he created mathematical equations to describe the behavior of light when it strikes a mirror, a lens, or a prism.

## Reflection and Refraction

*How Can You Predict Where a Ray of Light Will Go When It Comes in Contact With Different Types of Transparent Materials?*

## The Task

Use what you know about light, uncovering patterns in nature, and the use of models in science to design and carry out an investigation using a simulation to determine how light behaves when it travels through one transparent material and then enters into a different one.

The guiding question of this investigation is, **How can you predict where a ray of light will go when it comes in contact with different types of transparent materials?**

## Materials

You will use an online simulation called *Bending Light* to conduct your investigation. You can access the simulation by going to the following website: *https://phet.colorado.edu/en/simulation/bending-light*.

## Safety Precautions

Follow all normal lab safety rules.

## Investigation Proposal Required?  ☐ Yes  ☐ No

## Getting Started

The *Bending Light* simulation (see Figure L20.4, p. 194) enables you to change the angle of incidence of a light ray that crosses the boundary between two transparent materials and then measure the angle of reflection and refraction. You can also adjust the properties of the two materials and measure the light intensity of each light ray. To use this simulation, start by clicking on the "Intro" button. You will then see a laser pointer and a horizontal line that represents the boundary between two different materials. Click on the red button on the laser pointer to turn it on. This will allow you see a light ray and what happens to it as it crosses the boundary between the two transparent materials. You can change the angle of incidence of the light ray by clicking and dragging on the left end of the laser pointer. To measure the angle of incidence, the angle of reflection, and the angle of refraction, simply drag the protractor in the lower-left corner and drop it on the surface normal (which is represented by the dashed line). You can change the properties of the two transparent materials using the gray boxes on the right side of the screen. Finally, you can measure the light intensity of any ray by dragging and dropping the green light intensity meter where you need it. The green light intensity meter is located in the lower-left corner of the simulation.

## FIGURE L20.3

**A ray of light crossing the boundary between two transparent materials (air and plastic)**

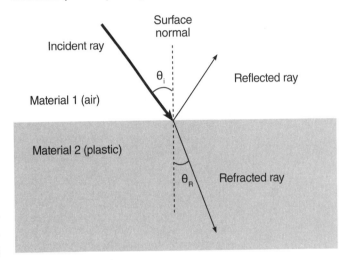

## FIGURE L20.4

**A screen shot of the *Bending Light* simulation**

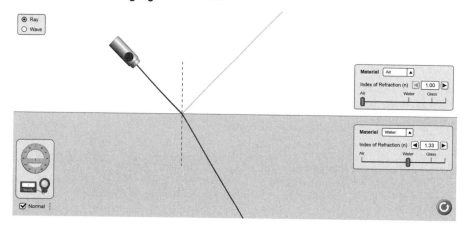

To answer the guiding question, you must determine what type of data you need to collect, how you will collect the data, and how you will analyze it. To determine *what type of data you need to collect,* think about the following questions:

- Which factors will you need to account for to be able to make accurate predictions?
- What type of measurements will you need to record?

To determine *how you will collect the data using the simulation,* think about the following questions:

- What will serve as your dependent variable or variables?
- What will serve as your independent variable or variables?
- How will you vary the independent variable?
- What will you do to hold the other variables constant during each experiment?
- What types of comparisons will you need to make using the simulation?
- How many comparisons will you need to make to determine a trend or a relationship?
- How will you keep track of the data you collect and how will you organize it?

To determine *how you will analyze the data,* think about the following questions:

- What type of calculations will you need to make?
- What type of graph could you create to help make sense of your data?

## Reflection and Refraction

*How Can You Predict Where a Ray of Light Will Go When It Comes in Contact With Different Types of Transparent Materials?*

Once you have collected the data you need, your group will need to use your findings to develop an answer to the guiding question for this investigation. Your answer to the guiding question must explain how to predict the path of the ray as it crosses the boundary between two transparent materials. For your claim to be sufficient, your answer will need to include both the angle of reflection and the angle of refraction. You can then transform the data you collected using the simulation to support the validity of your overall explanation.

### Connections to Crosscutting Concepts, the Nature of Science, and the Nature of Scientific Inquiry

As you work through your investigation, be sure to think about

- the importance of looking for and understanding patterns in data,
- the importance of using models to study natural phenomena in science,
- how scientific knowledge can change over time, and
- the culture of science and how it influences the work of scientists.

### Initial Argument

Once your group has finished collecting and analyzing your data, your group will need to develop an initial argument. Your initial argument needs to include a *claim*, *evidence* to support your claim, and a *justification* of the evidence. The claim is your group's answer to the guiding question. The evidence is an analysis and interpretation of your data. Finally, the justification of the evidence is why your group thinks the evidence matters. The justification of the evidence is important because scientists can use different kinds of evidence to support their claims. Your group will create your initial argument on a whiteboard. Your whiteboard should include all the information shown in Figure L20.5.

### Argumentation Session

The argumentation session allows all of the groups to share their arguments. One member of each group will stay at the lab station to share that group's argument, while the other members of the group go to the other lab stations to listen to and critique the arguments developed by their classmates. This is similar to how scientists present their arguments to other scientists at conferences. If you are responsible for critiquing your classmates' arguments, your goal is to look for mistakes so these mistakes can be fixed and they can make their argument better. The argumentation session is also a good time to think about ways you can make your initial argument better. Scientists must share and critique arguments like this to develop new ideas.

**FIGURE L20.5** _____

**Argument presentation on a whiteboard**

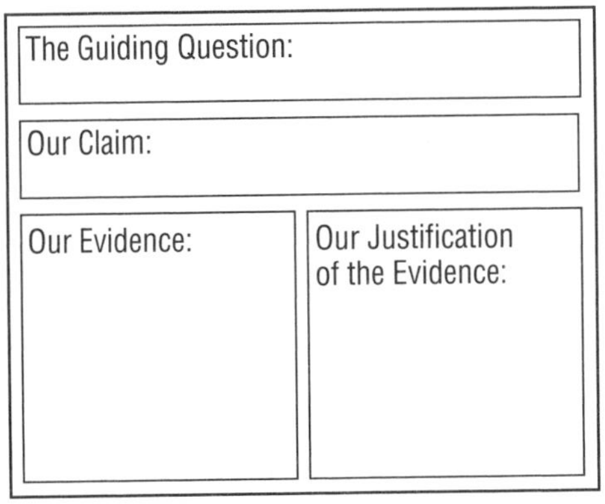

| The Guiding Question: | |
|---|---|
| Our Claim: | |
| Our Evidence: | Our Justification of the Evidence: |

To critique an argument, you might need more information than what is included on the whiteboard. You will therefore need to ask the presenter lots of questions. Here are some good questions to ask:

- How did you collect your data? Why did you use that method? Why did you collect those data?

- What did you do to make sure the data you collected are reliable? What did you do to decrease measurement error?

- How did your group analyze the data? Why did you decide to do it that way? Did you check your calculations?

- Is that the only way to interpret the results of your analysis? How do you know that your interpretation of your analysis is appropriate?

- Why did your group decide to present your evidence in that way?

- What other claims did your group discuss before you decided on that one? Why did your group abandon those alternative ideas?

- How confident are you that your claim is valid? What could you do to increase your confidence?

Once the argumentation session is complete, you will have a chance to meet with your group and revise your initial argument. Your group might need to gather more data or design a way to test one or more alternative claims as part of this process. Remember, your goal at this stage of the investigation is to develop the most acceptable and valid answer to the research question!

## Report

Once you have completed your research, you will need to prepare an *investigation report* that consists of three sections. Each section should provide an answer to the following questions:

1. What question were you trying to answer and why?

2. What did you do to answer your question and why?

3. What is your argument?

Your report should answer these questions in two pages or less. This report must be typed, and any diagrams, figures, or tables should be embedded into the document. Be sure to write in a persuasive style; you are trying to convince others that your claim is acceptable and valid!

**Reflection and Refraction**

*How Can You Predict Where a Ray of Light Will Go When It Comes in Contact With Different Types of Transparent Materials?*

## *Checkout Questions*

# Lab 20. Reflection and Refraction

## How Can You Predict Where a Ray of Light Will Go When It Comes in Contact With Different Types of Transparent Materials?

1.  A student is conducting an investigation in which she wants to shine a laser pointer into a tank of water so that the beam of light hits the center of a target on the bottom of the tank. Which position for the laser pointer gives her the best chance of hitting the center of the target: position A, position B, or position C?

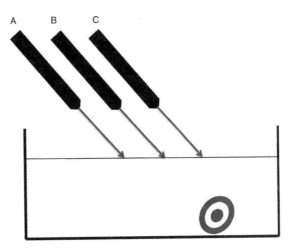

Explain why you chose that position as the best choice for the beam of light to hit the center of the target.

2. An engineer is designing a piece of equipment that will help change the path of a beam of light. The new piece of equipment needs to change the path of the light as much as possible. The table below has all of the materials, and the index of refraction for each one, that the engineer can choose from to help change the path of the light. Which material is the best choice?

| Material | Index of refraction |
| --- | --- |
| Crown glass | 1.52 |
| Ice | 1.31 |
| Pyrex glass | 1.47 |
| Clear plastic | 1.60 |
| Liquid water | 1.33 |

Explain why you chose that material as the best choice for the piece of equipment.

3. Once scientists learn about something new, their ideas do not change.

 a. I agree with this statement.

 b. I disagree with this statement.

Explain your answer, using an example from your investigation about reflection and refraction.

**Reflection and Refraction**

*How Can You Predict Where a Ray of Light Will Go When It Comes in Contact With Different Types of Transparent Materials?*

4. Scientists are objective, so they are not influenced by the culture of society.

    a.  I agree with this statement.

    b.  I disagree with this statement.

Explain your answer, using an example from your investigation about reflection and refraction.

5. Scientists study patterns in nature and patterns within the data that they collect. Explain why it is important for scientists to understand patterns, using an example from your investigation about reflection and refraction.

6. Models are useful tools that help scientists better understand what they are studying. These models can be conceptual, mathematical (such as equations or relationships), or physical (such as a drawing). Using an example from your investigation about reflection and refraction, explain why it is important for scientists to develop and use models in their work.

# Application Labs

# LAB 21

# Lab 21. Light and Information Transfer
## How Does the Type of Material Affect the Amount of Light That Is Lost When Light Waves Travel Down a Tube?

## Introduction

Starting in the late 1940s, scientists and mathematicians began conducting experiments that led to a new field of study that today we call information theory. Scientists and mathematicians who conduct research in the field of information theory focus on answering a few important questions. The first of these questions is, how can we transfer information from one place to another? By information transfer, scientists and mathematicians mean how information is shared between people and things. For example, you might have watched a sporting event over the weekend and know who won the game, while your friend was unable to watch and does not know who won. If you tell your friend who won, that is transferring information from you to your friend. Although not a new question (humans have been transferring information for thousands of years), the formal study of this question is quite new. Telephones, fax machines, and even the internet grew out of this type of research. Furthermore, many of the cables that you connect to your TV or computer serve the purpose of transferring information from someplace else to your TV or computer.

A second question that scientist and mathematicians who study information theory ask is, what are the advantages and disadvantages to transferring information in different ways? For example, the oldest ways to transfer information from one person to another person is by talking to that person. The advantages of this type of information transfer are that (1) it happens very quickly and (2) you know who sent the message because you can see him or her in front of you. The disadvantage is that the message does not last very long. Another way to transfer information is by writing a letter. The advantage of the letter is that it lasts a long time. The disadvantage of the letter is that it also takes a long time to mail a letter to a friend.

Related to the question about the advantages and disadvantages of information transfer is the question, how can we get a message to another person in the least amount of time possible? Scientists who study physics (another branch of science) have determined that light moves faster than anything else in the universe. Information scientists used this finding to answer their question about transmitting a message as fast as possible. If light is the fastest thing in the universe, then maybe light can be used to transfer information.

Another question that information scientists ask about information transfer is, how can we limit the loss of information when transferring it from one place to another? Information scientists have determined that all messages lose some information between being sent and being received. Sometimes this is not a problem; for example, if you write a letter to a friend, the letter will not transfer information about whether you wrote it while sitting inside or wrote it while sitting outside (unless you say so in the letter). Other times, however, loss of information is a problem. If you have ever heard static while you talked to another person on the phone, this is an example of information loss.

## Your Task

Use what you know about light, tracking energy and matter, and the relationship between structure and function to design and carry out an investigation that will allow you to determine how much light is lost when it shines down different types of tubing. This will allow you to make a recommendation about what type of materials we should use for transferring information with light. It is also important to recognize that many of the cables we use to transfer information are able to bend, so that we can get the cables to go in whatever direction we want. To complete this task, you will need to test how much light makes it from one end of a tube to the other end, when the tube has a 45° bend in the middle.

The guiding question of this investigation is, **How does the type of material affect the amount of light that is lost when light waves travel down a tube?**

## Materials

You may use any of the following materials during your investigation:

- Light source
- Light sensor with interface
- Electrical tape
- Fiber optic tubing
- Amber rubber tubing
- Red vacuum and pressure tubing
- Tygon laboratory tubing
- Protractor
- Safety glasses or goggles

## Safety Precautions

Follow all normal lab safety rules. In addition, take the following safety precautions:

1. Wear sanitized safety glasses or goggles during lab setup, hands-on activity, and takedown.

2. Use caution when working with the light source, because it can get hot and burn skin.

3. Use only GFCI-protected electrical receptacles for the lamp power source.

4. Do not shine a laser pointer at anyone's eyes or face.

5. Lightbulbs are made of glass. Be careful handling them. If they break, clean them up immediately and place in a broken glass box.

6. Wash hands with soap and water after completing the lab activity.

# LAB 21

**Investigation Proposal Required?** ☐ Yes ☐ No

## Getting Started

To answer the guiding question, you will need to design and conduct an investigation to measure the amount of light that is lost when it shines down a tube. To accomplish this task, you must determine what type of data you need to collect, how you will collect it, and how you will analyze it before you begin.

To determine *what type of data you need to collect,* think about the following questions:

- How will you determine how much light enters the tube?
- How will you determine how much light exits the tube?

To determine *how you will collect your data,* think about the following questions:

- What equipment will you need to collect the data you need?
- How will you make sure that your data are of high quality (i.e., how will you reduce error)?
- Are there different ways you can measure the amount of light transferred?
- How will you keep track of the data you collect?
- How will you organize your data?

To determine *how you will analyze your data,* think about the following questions:

- How will you determine the amount of light lost?
- What type of table or graph could you create to help make sense of your data?

## Connections to Crosscutting Concepts, the Nature of Science, and the Nature of Scientific Inquiry

As you work through your investigation, be sure to think about

- the importance of tracking how energy and matter move into, out of, and within systems;
- the relationship between structure and function in nature;
- science as a culture and how it influences the work of scientists; and
- the importance of imagination and creativity in science.

## Initial Argument

Once your group has finished collecting and analyzing your data, your group will need to develop an initial argument. Your initial argument needs to include a *claim, evidence* to support your claim, and a *justification* of the evidence. The claim is your group's answer to

National Science Teachers Association

the guiding question. The evidence is an analysis and interpretation of your data. Finally, the justification of the evidence is why your group thinks the evidence matters. The justification of the evidence is important because scientists can use different kinds of evidence to support their claims. Your group will create your initial argument on a whiteboard. Your whiteboard should include all the information shown in Figure L21.1.

## FIGURE L21.1

**Argument presentation on a whiteboard**

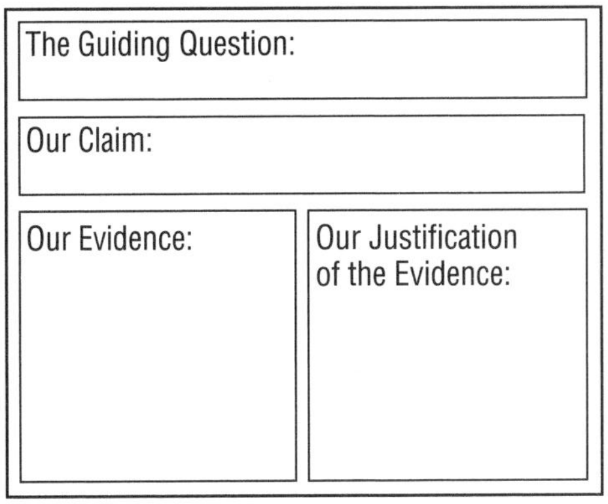

| The Guiding Question: | |
| --- | --- |
| Our Claim: | |
| Our Evidence: | Our Justification of the Evidence: |

### Argumentation Session

The argumentation session allows all of the groups to share their arguments. One member of each group will stay at the lab station to share that group's argument, while the other members of the group go to the other lab stations to listen to and critique the arguments developed by their classmates. This is similar to how scientists present their arguments to other scientists at conferences. If you are responsible for critiquing your classmates' arguments, your goal is to look for mistakes so these mistakes can be fixed and they can make their argument better. The argumentation session is also a good time to think about ways you can make your initial argument better. Scientists must share and critique arguments like this to develop new ideas.

To critique an argument, you might need more information than what is included on the whiteboard. You will therefore need to ask the presenter lots of questions. Here are some good questions to ask:

- How did you collect your data? Why did you use that method? Why did you collect those data?
- What did you do to make sure the data you collected are reliable? What did you do to decrease measurement error?
- How did your group analyze the data? Why did you decide to do it that way? Did you check your calculations?
- Is that the only way to interpret the results of your analysis? How do you know that your interpretation of your analysis is appropriate?
- Why did your group decide to present your evidence in that way?
- What other claims did your group discuss before you decided on that one? Why did your group abandon those alternative ideas?
- How confident are you that your claim is valid? What could you do to increase your confidence?

Once the argumentation session is complete, you will have a chance to meet with your group and revise your initial argument. Your group might need to gather more data or design a way to test one or more alternative claims as part of this process. Remember, your goal at this stage of the investigation is to develop the most acceptable and valid answer to the research question!

## Report

Once you have completed your research, you will need to prepare an *investigation report* that consists of three sections. Each section should provide an answer to the following questions:

1.  What question were you trying to answer and why?

2.  What did you do to answer your question and why?

3.  What is your argument?

Your report should answer these questions in two pages or less. This report must be typed, and any diagrams, figures, or tables should be embedded into the document. Be sure to write in a persuasive style; you are trying to convince others that your claim is acceptable and valid!

*Checkout Questions*

# Lab 21. Light and Information Transfer
## How Does the Type of Material Affect the Amount of Light That Is Lost When Light Waves Travel Down a Tube?

1. A student shines a flashlight on a glass window from across the room. Using arrows, draw the different paths that the beam of light might take when they reach the window.

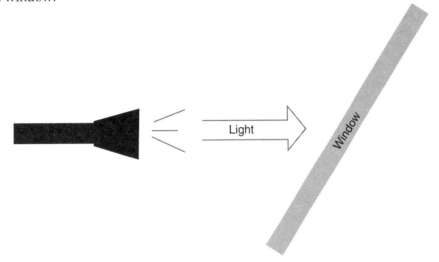

Explain why you drew the arrows in those directions.

# LAB 21

2. Two engineers are trying to decide how to send information signals and messages between two different locations. One engineer wants to use electrical cables to send the signals and messages because that technique has been used for a very long time. The other engineer wants to use fiber optic cables and the newer technique of sending signals and messages with light. Compare and contrast the two approaches, and make a recommendation to the engineers on which approach they should take to send their signals and messages.

3. Current events influence the research that scientist do.

    a. I agree with this statement.
    b. I disagree with this statement.

    Explain your answer, using an example from your investigation on light and information transfer.

4. Imagination and creativity are traits that only artists need, not scientists.

    a.  I agree with this statement.
    b.  I disagree with this statement.

    Explain your answer, using an example from your investigation on light and information transfer.

5. Solving problems in science or engineering often means developing a new structure, tool, or piece of equipment. Each of these objects will have a specific function that helps solve the problem. Explain why it is important for scientists and engineers to understand the relationship between a structure and its function, using an example from your investigation about light and information transfer.

6. Two of the most important laws related to the natural world are the law of conservation of matter and the law of conservation of energy. Using an example from your investigation on light and information transfer, explain why it is important for scientists to understand and track the flow of matter and energy into, out of, and within systems.

## Lab Handout

# Lab 22. Design Challenge
## How Should Eyeglasses Be Shaped to Correct for Nearsightedness and Farsightedness?

### Introduction

The study of light, an area in physics known as optics, dates back to the times of the ancient Mesopotamians, Egyptians, Greeks, and Romans. It is believed that the first lenses were made as early as 750 B.C. (see the Nimrud lens at *www.britishmuseum.org/research/collection_online/ collection_object_details.aspx?objectId=369215&partId=1&searchText=lens&page=1*). Early lenses were used to manipulate light and likely most often used to start fires by focusing light in a small area to generate enough heat to ignite flammable material. Over time, scientists have used their understanding of the properties of light to develop many useful instruments for their investigations and for society, including telescopes, microscopes, magnifying glasses, and eyeglasses. Each of these instruments uses at least one lens to change the path of light rays to be more beneficial to the person using the instrument.

Light rays behave in predictable ways. There are three general ways that light rays can behave when they interact with, or pass through, a lens: they can be reflected, transmitted, or absorbed. When light rays are reflected, that means they come into contact with a surface and bounce back in the direction they came from; this is called reflection. When light rays come into contact with a surface and continue on, passing through the surface, it is called transmission. The third behavior for light rays is that when they hit a surface they may not be reflected or transmitted but instead are absorbed. In many cases when light rays hit a surface, a combination of these behaviors happens. For example, some rays may get reflected while others are transmitted. Also, when light rays are reflected or transmitted, it is common for them to change direction. When light rays are transmitted through a substance but change direction on the other side, it is called refraction. Figure L22.1 shows examples of what happens when light rays are reflected, transmitted, or absorbed.

When light rays are transmitted through an object, such as a lens, the light is refracted in specific ways based on the shape of the object. Scientists have conducted many investigations to understand how light rays behave when they pass through a lens. Two major

## FIGURE L22.1 _____

Examples of how light behaves when it interacts with a medium: (a) reflection, (b) transmission, (c) transmission with refraction, and (d) absorption. The arrows represent light rays.

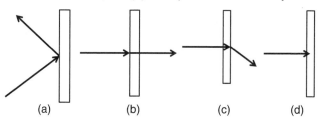

findings from these investigations are as follows: When light rays are transmitted through a convex lens, the light rays come together, or converge on the other side; when light rays are transmitted through a concave lens, the light rays spread out, or diverge, on the other side. Figure L22.2 shows how light rays behave when they pass through a convex or concave lens. Glass lenses used in instruments like the ones described earlier in this section are very common, and even our own eyes have lenses that collects light rays to help us see.

# FIGURE L22.2

**A convex (a) and a concave (b) lens refracting light**

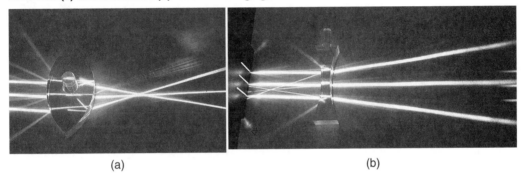

(a)　　　　　　　　　　　　　　　　　(b)

The lens in a human eye focuses the incoming light rays on the retina, which is the back portion of the eyeball. However, there are times when a person's eye does not focus the light correctly, resulting in the person being nearsighted or farsighted. When a person is nearsighted, the lenses in the eyes focus the incoming light rays before they have a chance to reach the retina. When a person is farsighted, the lenses in the eyes do not focus the incoming light rays fast enough and they are still spread out when the light rays reach the retina.

Eyeglasses or contact lenses are used to correct the vision of people with nearsightedness or farsightedness. It is believed that eyeglasses were first invented in the 1200s and then gained popularity in the mid-1400s with the invention of the printing press and the rise in the number of people that had access to books and began learning to read (see "Timeline of Eyeglasses" at *www.museumofvision.org/exhibitions/?key=44&subkey=4&relkey=35*). The lenses of the eyeglasses (or modern contact lenses) work together with the lenses of the eye to change the path of the incoming light rays to ensure that they focus on the retina, resulting in clear vision. Figure L22.3 (p. 212) shows examples of eyes and incoming light rays that represent normal vision, nearsightedness, and farsightedness.

# LAB 22

FIGURE L22.3

**Eyeball models for (a) normal vision, (b) nearsightedness, and (c) farsightedness**

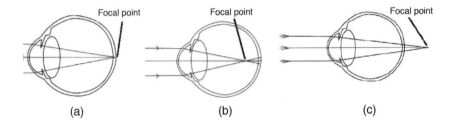

(a)                    (b)                    (c)

## Your Task

Use what you know about the behavior of light and the relationship between the structure and function of a lens to develop a model that helps you explain how different shapes of eyeglasses will correct the vision of someone who is nearsighted and someone who is farsighted. Your model should demonstrate the two types of vision conditions as well as show how your solution corrects each of the vision conditions.

The guiding question of this investigation is, **How should eyeglasses be shaped to correct for nearsightedness and farsightedness?**

## Materials

You may use any of the following materials during you investigation:

**Consumable**
Large paper

**Equipment**
- Light box with power supply
- Lens kit
- Protractor
- Ruler
- Safety glasses or goggles

## Safety Precautions

Follow all normal lab safety rules. In addition, take the following safety precautions:

1. Wear sanitized safety glasses or goggles during lab setup, hands-on activity, and takedown.

2. Use caution when working with the light source. It can get hot and burn skin.

3. Do not look directly at the light coming from the light box.

4. Use only GFCI-protected electrical receptacles for the light box power supply.

5. Use caution when handling glass. It can have sharp edges, which can cut skin.

6. Lightbulbs are made of glass. Be careful handling them. If they break, clean them up immediately and place in a broken glass box.

7. Wash hands with soap and water after completing the lab activity.

**Investigation Proposal Required?**  ☐ Yes    ☐ No

## Getting Started

The first step in developing your vision models is to use lenses from your kit to determine how to draw a model eyeball that represents normal vision. Then, draw a model eyeball that represents nearsightedness and a model eyeball that represents farsightedness. Remember that each model eyeball needs a lens that remains part of the model eyeball (this lens represents the lens portion of the human eye, and removing it would be the same as doing surgery on your model eyeball), and that lens cannot be used to represent the eyeglasses intended to correct the vision in that model. Work with each model separately; this allows you to use a lens from your kit more than once if necessary.

To determine *what type of data you need to collect,* think about the following questions:

- What information do you need to make your models?
- What measurements will you take during your investigation?
- How will you know if the vision has been corrected in your models?

To determine *how you will collect your data,* think about the following questions:

- What equipment will you need to collect the data you need?
- How will you make sure that your data are of high quality (i.e., how will you reduce error)?
- How will you keep track of the data you collect?
- How will you organize your data?

To determine *how you will analyze your data,* think about the following questions:

- How can you show that the vision or light rays in your model have changed?
- What type of diagrams or images could you create to help make sense of your data?

## Connections to Crosscutting Concepts, the Nature of Science, and the Nature of Scientific Inquiry

As you work through your investigation, be sure to think about

- how scientists use models to understand complex systems;

# LAB 22

- how the structure and function of an object are related;

- science as a culture and how it influences the work of scientists; and

- how scientists must use imagination and creativity when developing models and explanations.

## Initial Argument

Once your group has finished collecting and analyzing your data, your group will need to develop an initial argument. Your initial argument needs to include a *claim, evidence* to support your claim, and a *justification* of the evidence. The claim is your group's answer to the guiding question. The evidence is an analysis and interpretation of your data. Finally, the justification of the evidence is why your group thinks the evidence matters.

## FIGURE L22.4

**Argument presentation on a whiteboard**

| The Guiding Question: | |
|---|---|
| Our Claim: | |
| Our Evidence: | Our Justification of the Evidence: |

The justification of the evidence is important because scientists can use different kinds of evidence to support their claims. Your group will create your initial argument on a whiteboard. Your whiteboard should include all the information shown in Figure L22.4.

## Argumentation Session

The argumentation session allows all of the groups to share their arguments. One member of each group will stay at the lab station to share that group's argument, while the other members of the group go to the other lab stations to listen to and critique the arguments developed by their classmates. This is similar to how scientists present their arguments to other scientists at conferences. If you are responsible for critiquing your classmates' arguments, your goal is to look for mistakes so these mistakes can be fixed and they can make their argument better. The argumentation session is also a good time to think about ways you can make your initial argument better. Scientists must share and critique arguments like this to develop new ideas.

To critique an argument, you might need more information than what is included on the whiteboard. You will therefore need to ask the presenter lots of questions. Here are some good questions to ask:

- How did you collect your data? Why did you use that method? Why did you collect those data?

- What did you do to make sure the data you collected are reliable? What did you do to decrease measurement error?

- How did you group analyze the data? Why did you decide to do it that way? Did you check your calculations?

- Is that the only way to interpret the results of your analysis? How do you know that your interpretation of your analysis is appropriate?

- Why did your group decide to present your evidence in that way?

- What other claims did your group discuss before you decided on that one? Why did your group abandon those alternative ideas?

- How confident are you that your claim is valid? What could you do to increase your confidence?

Once the argumentation session is complete, you will have a chance to meet with your group and revise your initial argument. Your group might need to gather more data or design a way to test one or more alternative claims as part of this process. Remember, your goal at this stage of the investigation is to develop the most acceptable and valid answer to the research question!

## Report

Once you have completed your research, you will need to prepare an *investigation report* that consists of three sections. Each section should provide an answer to the following questions:

1. What question were you trying to answer and why?

2. What did you do to answer your question and why?

3. What is your argument?

Your report should answer these questions in two pages or less. This report must be typed, and any diagrams, figures, or tables should be embedded into the document. Be sure to write in a persuasive style; you are trying to convince others that your claim is acceptable and valid!

## *Checkout Questions*

# Lab 22. Design Challenge
## How Should Eyeglasses Be Shaped to Correct for Nearsightedness and Farsightedness?

1. Below are two different convex lenses. Show how the path of the light rays would change when they pass through each lens. The arrows represent the incoming light rays.

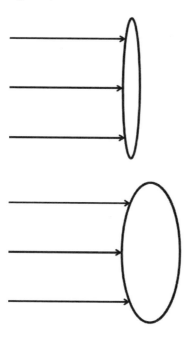

Explain why you chose those paths for the light rays for each lens.

2. Eyeglasses are worn close to your eye. Use what you know about concave and convex lenses and the behavior of light to describe how the curvature of lenses in eyeglasses would need to change to maintain clear vision if they were worn farther away from your eye.

3. Scientists only study topics that have direct application to society.

   a. I agree with this statement.
   b. I disagree with this statement.

   Explain your answer, using an example from your investigation about eyeglasses and the behavior of light.

4. Imagination and creativity are important characteristics for scientists.

   a. I agree with this statement.
   b. I disagree with this statement.

   Explain your answer, using an example from your investigation about eyeglasses and the behavior of light.

5. When the printing press became popular, the number of people reading books increased, and this is about the same time that eyeglasses started to gain popularity as a means to correct blurry vision. Eyeglasses have a specific structure and serve a distinct purpose. Explain why it is important for scientists and engineers to understand the relationship between a structure and its function, using an example from your investigation about eyeglasses and the behavior of light.

6. Scientists and engineers often make models to help them understand complex systems and phenomena. These models can be conceptual, mathematical, or physical—like a drawing. Using an example from your investigation about eyeglasses and the behavior of light, explain why it is important for scientists and engineers to develop models in their work.

# IMAGE CREDITS

## Lab 1

Figure L1.1: Authors

Figure L1.2: Authors

Figure L1.3: Authors

Checkout Questions figures: Authors

## Lab 2

Figure L2.2: Authors

## Lab 3

Figure L3.1: Authors

Figure L3.2: Authors

## Lab 4

Figure L4.1: Authors

Figure L4.2: Authors

Checkout Questions figures: Authors

## Lab 5

Figure L5.1: Authors

Figure L5.2: Authors

Checkout Questions figures: Vacuum-style container: User:Dhscommtech, Wikimedia Commons, CC BY-SA 3.0, GFDL 1.2. *https://commons.wikimedia.org/wiki/File:Thermos.JPG*; Cross-section of a vacuum-style container: Authors.

## Lab 6

Figure L6.1: Adapted from *http://manashsubhaditya.blogspot.com/2012/06/gravity-and-upthurst-two-opposite.html.*

Figure L6.2: Wikimedia Commons, Public domain. *https://commons.wikimedia.org/wiki/File:Nikolaus_Kopernikus.jpg*

Figure L6.3: PhET Interactive Simulations, University of Colorado Boulder *http://phet.colorado.edu*; *http://phet.colorado.edu/en/simulation/gravity-force-lab*

Figure L6.4: Authors

Checkout Questions figures: Authors

## Lab 7

Figure L7.2: Authors

## Lab 8

Figure L8.1: User:Cdang, Wikimedia Commons, Public domain. *https://commons.wikimedia.org/wiki/File:Tir_a_la_corde_equilibre.svg*

Figure L8.2: Authors

Figure L8.3: Authors

Checkout Questions figure: Authors

## Lab 9

Figure L9.1: GOGO Visual, Wikimedia Commons, CC BY 2.0. *https://commons.wikimedia.org/wiki/File:Toni_El%C3%ADas.jpg*

Figure L9.2: Authors

Figure L9.3: Authors

## Lab 10

Figure L10.1: Adapted from Victor Blacus, Wikimedia Commons, CC BY-SA 3.0. *https://commons. wikimedia.org/wiki/File:Magnetic_field_of_a_steady_ current.svg*

Figure L10.2: Adapted from Zuriks, Wikimedia Commons, Public domain. *https://commons.wikimedia. org/wiki/File:Solenoid-1_%28vertical%29.png*

Figure L10.3: Authors

Figure L10.4: Authors

Checkout Questions figure: Authors

## Lab 11

Figure L11.1: Authors

Figure L11.2: Authors

Checkout Questions figures: Authors

## Lab 12

Figure L12.1: Authors

Figure L12.2: Authors

## Lab 13

Figure L13.2: Authors

Checkout Questions figure: Authors

## Lab 14

Figure L14.1: User:The light and the dark, Wikimedia Commons, CC BY-SA 3.0, GFDL 1.2. *https://commons.wikimedia.org/wiki/File:Korean_ teeterboard.JPG*

Figure L14.2: Authors

Checkout Questions figure: Authors

## Lab 15

Figure L15.1: Authors

Figure L15.2: Authors

Figure L15.3: Authors

Checkout Questions figure: Authors

## Lab 16

Figure L16.1: Authors

Figure L16.2: Authors

Checkout Questions figure: Authors

## Lab 17

Figure L17.1: Authors

Figure L17.2: Authors

Checkout Questions figures: Authors

## Lab 18

Figure L18.1: Jonathan S Urie, Wikimedia Commons, CC BY-SA 3.0. *https://commons.wikimedia.org/ wiki/Category:Electromagnetic_spectrum#/media/ File:BW_EM_spectrum.png*

Figure L18.2: Authors

Figure L18.3: Authors

Checkout Questions figure: Authors

## Lab 19

Figure L19.2: PhET Interactive Simulations, University of Colorado Boulder *http://phet.colorado.edu*; *https://phet.colorado. edu/sims/html/wave-on-a-string/latest/wave-on-a- string_en.html*

Figure L19.3: Authors

Checkout Questions figures: Authors

## Lab 20

Figure L20.2: Modified from Philip Ronan, Wikimedia Commons, GFDL 1.2. *https://en.wikipedia.org/wiki/ Frequency#/media/File:EM_spectrum.svg*

Figure L20.3: Authors

Figure L20.4: PhET Interactive Simulations, University of Colorado Boulder *http://phet.colorado.edu*; *https://phet.colorado.edu/ en/simulation/bending-light*

Figure L20.5: Authors

Checkout Questions figure: Authors

## Lab 21

Figure 21.1: Authors

Figure 21.2: Authors

Figure 21.3: Authors

Figure L21.1: Authors

Checkout Questions figure: Authors

## Lab 22

Figure L22.1: Authors

Figure L22.2: a: User:Leridant, Wikimedia Commons, CC BY-SA 3.0, GFDL 1.2. *https://commons. wikimedia.org/wiki/File:Large_convex_lens.jpg*; b: User:Leridant, Wikimedia Commons, CC BY-SA 3.0, GFDL 1.2. *https://en.wikipedia.org/wiki/ File:Concave_lens.jpg*

Figure L22.3: a: Adapted from CryptWizard, Wikimedia Commons, CC SA 1.0. *http://en.wikipedia. org/wiki/File:Hypermetropia.svg*; b: Adapted from A. Baris Toprak, Wikimedia Commons, CC SA 1.0. *http:// en.wikipedia.org/wiki/File:Myopia.svg*; c: Adapted from CryptWizard, Wikimedia Commons, CC SA 1.0. *http://en.wikipedia.org/wiki/File:Hypermetropia.svg*

Figure L22.4: Authors

Checkout Questions figure: Authors